You Can Farm

the entrepreneur's guide to start and succeed in a farm enterprise

by

Joel Salatin

Polyface, Inc.
Swoope, Virginia

Other books

by

Joel Salatin

Pastured Poultry Profit$
Salad Bar Beef
Family Friendly Farming
Holy Cows & Hog Heaven
Everything I Want to do is Illegal

All books available from:

Chelsea Green Publishing	1-800-639-4099
Amazon.com	www.amazon.com
ACRES USA	1-800-355-5313
Stockman Grass Farmer	1-800-748-9808
Back 40 Books	1-573-858-3559
Cumberland Books	1-606-787-4089

or by special order from your local bookstore.

This publication is designed to provide accurate and authoritative information in regard to the subject matter covered. It is sold with the understanding that the publisher is not engaged in rendering legal, accounting or other professional service. If legal advice or other expert assistance is required, the services of a competent professional person should be sought. -- *From a declaration of principles jointly adopted by a committee of the American Bar Association and a committee of publishers.*

Editing and layout by Vicki Dunaway

About the cover: *Profit, Production and Pleasure all come together in this early spring photograph at Polyface Farm in Virginia's Shenandoah Valley. In the foreground a "Harepen" houses pastured rabbits. "Salad Bar Beef" cattle prepare the grass for pastured broilers in their protective portable pens. To the right, laying hens from the "Eggmobile" sanitize the paddock vacated by the cows. Grassland, forest land and water meld into a harmonious landscape pattern.*

Library of Congress Catalog Card Number: 98-066457

ISBN: 0-9638109-2-8

Contents

Contents (continued)

Acknowledgements

Because this book is an outgrowth of our family's personal success at farming, I must thank my wife, Teresa, first and foremost for forgoing fancy cars, fancy furniture and fancy vacations as, together, we sacrificed for a dream. She is more frugal than I, and that has made all the difference.

Secondly, that this book can be written at this stage in my life is a direct result of having a running start compliments of Dad and Mom, who did not accumulate debt during their years of part-time farming. Their leaving Teresa and me with a gift of land — and the philosophy to capitalize on it — vaulted us ahead of the average aspiring farm family.

Our children, Daniel and Rachel, exhibit the same characteristics necessary for success, and are now building their own farm enterprises. Teresa and I trust that they will go far beyond us, as they should. Our apprentices, especially Josh Griggs and Joel Bauman, picked up the slack during the winter, enabling me to spend days at the computer composing this volume.

Many thanks, as usual, go to Vicki Dunaway, whose special talents for desktop publishing and critical reading eye pick up where I leave off. She gladly does the indexing and pulls all the loose ends together to make the manuscript a real book.

Finally, all of our farm patrons, all the wonderful folks in our fellowship group community, and all my creative farming friends around the country deserve accolades for giving me new insights and encouraging me to pursue this project. God bless each of you.

Foreword

Do you have a dream of living on a farm? A great many people do and with good reason. It is the ultimate family business and one where you can share your daily work experience with even the smallest of children.

While it's true some farmers complain about the pay, virtually none of them want to give up the way of life. It is within this desire for a quality lifestyle that lies the great trap for the unwary. Living on a farm and making a living from a farm are two entirely different things.

While it's true farming is about glorious sunsets, newborn lambs gamboling in the clover, frost on the pumpkin and snow in the pines at Christmas; it is also about money. It is about the harsh discipline of self-employment. It is about capital allocation and learning to not be cheap but "tight." It is about learning to hate things that rust, rot and lose value with time and use. It is about seeking and seizing opportunities. It is about understanding and exploiting the "unfair advantage" your climate, location and skills have given you. If these latter "business" things do not excite and intrigue you as much as that thought of the sunset, I recommend you don't try to farm for a living.

Conversely, if you have been a successful businessperson and want to change careers, you need to realize that dealing with a biological system requires far more patience and flexibility than a "normal" business. I have seen a lot of "type A" people beat themselves and their farms to death trying to make their farms do what they biologically couldn't do. Many new farms are lost because their computerized business plans did not have enough slack in them to survive a drought, a flood, or a hailstorm.

Also, you and your family should be prepared for the rather lengthy period of self-apprenticeship you will have to go through to learn the skills of a profitable farmer. Keep in mind that an apprentice's wages are far lower than those of a professional are, and a period of financial self-sacrifice will be required.

A good test of whether or not you are really cut out for the farm life is the attitude you have when you finish reading this book. Joel has done an excellent job of laying out both the good and the bad of the farm life. Joel and I discussed the fact that a great many people after having read this book would be dissuaded from trying to farm. We both agreed that if that was true the reader would never have made it as a farmer anyway and the price of this book would be the best investment he or she ever made.

I urge every member of your family who is going to be a part of your farm "team" to also read this book. It is extremely important that the expectations of your family be congruent with the reality that they are about to experience. A farm life can be a wonderful family experience. The best way to ensure it will be a successful experience is to go into it with as much knowledge as you can gather before you start farming. Reading this book is an excellent place to start.

Allan Nation, Editor
The Stockman Grass Farmer

Introduction

Have you ever desired, deep within your soul, to make a comfortable full-time living from a farming enterprise? Too often people dare not even vocalize this desire because it seems absurd. It's like thinking the unthinkable.

After all, the farm population is dwindling. It takes too much capital to start. The pay is too low. The working conditions are dusty, smelly and noisy: not the place to raise the family. This is all true, and more, for most farmers.

But for farm entrepreneurs, the opportunities for a farm family business have never been greater. The aging farm population is creating cavernous niches begging to be filled by creative visionaries who will go in dynamic new directions. As the industrial agriculture complex crumbles and our culture clambers for clean food, the countryside beckons anew with profitable farming opportunities.

While this book can be helpful to all farmers, it targets the wannabes, the folks who actually entertain notions of living, loving and learning on a piece of land. Anyone *willing* to dance with such a dream should be able to assess its assets and liabilities; its fantasies and realities. "Is it really possible for me?" is the burning question this book addresses.

Although the overall thrust is positive and exudes a "can do" spirit, the hurdles and misconceptions are treated honestly and seriously. As a fourth-generation farmer with a deep legacy of alternative thinkers on whose shoulders I stand, my advice may not be 100 percent correct, but it is tempered with lifetimes of wisdom. This practical, dirt-under-the-fingernails information is desperately needed as thousands of city folks look for land roots, seek something better than a cubicle at the end of the freeway, and begin repopulating the countryside dehumanized by industrial agriculture.

Nothing makes a competitor more successful than intimate knowledge of the opponent's strengths and weaknesses. My prayer is that ***You Can Farm*** will acquaint you with farming in ways you never imagined. It should stimulate thoughts you never had before. And it should awaken the dream within you in ways you never thought possible.

My goal is to empower you with new confidence, new vision and new direction. While you may not have some of the personal qualities necessary to embark on a farming career, a roadmap can make the journey far more possible. I trust that this book will inspire you with new hope. Let's look at the map together.

Joel Salatin
July 1998

Envisioning Your Future

Developing a Vision

The prerequisite to a successful farming enterprise is to believe it is possible. Every facet of our culture disparages the notion that farming can be enjoyable and economically viable.

I listened to a gifted public speaker in a 4-H contest once who recounted his run-ins with his high school guidance counselor when she found out he wanted to be a farmer. "Why would you want to throw your life away like that?" she queried, incredulously. "You have too much potential for that."

I can remember my own encounters with high school guidance counselors when I wanted to take typing instead of physics. After all, I was on the college fast track, the road to academia, and guidance counselors were supposed to protect gifted students from wasting their time on vocational choices like typing or agriculture classes.

The "poor, dumb farmer" stereotype receives encouragement from the comic strip *Garfield* every time Jon Arbuckle visits his parents. Any TV sitcom, if it mentions farming, does so with a smirk and some condescending, disparaging remark.

Unfortunately, the source for these ideas is the conventional paradigm, which indeed is a dead-end deal. But all over the country, individuals are discovering a totally different way to farm and they are doing it successfully.

The mainline model ends in despair and poverty. The alternative models can end in success and wealth. I am not talking necessarily about accumulating material wealth, although that is much of what this book is about. Right now we do not have people even thinking that subsistence farming is possible, much less really lucrative farming.

I've actually been publicly taken to task for even suggesting that it is possible to farm full-time in a way that is so enjoyable and lucrative that our children would actually want to do it. As long as in our heart of hearts we actually believe that it is impossible, we will never open up our spirits and our minds to ways to make it happen. We will dismiss even helpful hints, workable suggestions, with that dubious mindset that screens good ideas from our heads. "But what about?" permeates every sentence from naysayers and destroys all possibility of progress. A pessimistic outlook leaves no room for positive alternatives.

Allan Nation, editor of *Stockman Grass Farmer*, describes what we're up against with this little ditty: "It's easier to change the person than change the person's mind." I will not attempt to argue you into a vision. If you want to catch one, I can show you how to catch it. But this is not a book to debate you into a vision of successful farming.

A fellow came up to me after one of my conference presentations and said he found out there was a new rule among agriculture scientists who debunk the viability of small-scale, alternative farms: "They used to say 'I'll believe it when I see it.' Now they say 'I'll see it when I believe it.'"

In other words, your heart filters what your head can believe. If someone doesn't want to believe something, all the charts, graphs and argument in the world won't convince them. You have to be open to ideas before they can take any root in your head.

I will not attempt to convince naysayers. But folks who want to believe it's possible to farm need to be encouraged and shown the way. That is my goal.

Our culture lacks vision in almost every arena. How do you get vision? You have to be passionate about something. You must visualize the life you want to create and then be disciplined enough to get there. Really, accomplishing your dream is not so much about mechanics and opportunity as it is about character qualities: self-denial, perseverance, commitment, focus.

Hundreds of thousands of people dislike the notion of sitting in a cubicle in an office building at the end of the freeway punching some time clock for a nameless, faceless entity. Job security is a thing of the past. The information age is fragmenting our culture into cottage industries and designer anything. Outsourcing is the new buzzword among corporate empires.

Faith in the system is at an all-time low. Does anyone really trust Tyson to produce clean, nutritional chicken? Does anyone really trust the government to honestly inspect these plants? This lack of trust spawns brand new niche marketing opportunities for unique food. Public awareness, stimulated by media and the Internet, is creating a demand for products that would have had no public acceptance just two decades ago.

I remember when we were selling organic products at the local farmers' market in the late 1960s and early 1970s people didn't have a clue about sustainable and organic agriculture. Now everyone has heard of those terms and the niche is getting wider with each outbreak of food borne illness from the mega-processing facilities.

Another reason the opportunities are good now is because the farm population is aging and that means anyone with new information has an advantage in production. Because the older farmers are generally not open to new ideas, especially ideas as radical as the ones I propose in this book, you can be fairly assured that these successful models are not going to be adopted on a grand scale anytime in the near future.

Anyone who gets in on the front end now can command the lion's share of the marketplace because the majority of producers will continue down the self-destructive path outlined by the agribusiness community and USDA for several years. I represent the lunatic fringe of tomorrow's models. Who would have guessed 20 years ago, for example, that supermarkets would actually carry organic produce or Laura's Lean Beef? That was unthinkable. Consumer perception and producer dissatisfaction had not matured sufficiently to support such a scheme.

This highly successful marketing arrangement required both a public perception that mainline beef was unacceptable and enough farmers looking for something besides the sale barn to supply the product. Once disenchantment with the system included enough of both producers and consumers the logistical and economic viability of a mass marketed alternative product became possible. The tendency is, of course, for distinctiveness to drop as expansion continues, opening the way to new niches.

The next step in this progression is non-supermarket shopping as the information age continues to fragment industrialization in most of its manifestations. People are looking for designer anything, for uniqueness, for relationships. As we farmers concentrate on these needs, rather than emphasizing the same-old, same-old production, chemicals and efficiency, we will enjoy unprecedented opportunities.

Farmers' opportunities have dried up in direct proportion to how they have treated people ("bunch of ignorant city whiners"), how they have focused on production ("I don't care about quality"), how they have depended on chemicals ("got to kill those bugs"). That is why farming is in the doldrums and has a bad reputation. Farmers are harvesting a crop of bad philosophy, of misplaced emphasis.

With a 180-degree turn, however, all these negatives turn into positives, a special advantage to the enterprising alternative farmer. Just to prove what I am saying, walk up to the average farmer, or extension agent for that matter, and ask him if he's ever heard of *Stockman Grass Farmer* or *Acres USA* or Sir Albert Howard or Wendell Berry or Wes Jackson or Allan Savory and Holistic Management. You will probably get a blank stare.

But because you are aware of all these things, you suddenly

have a real information jump, a leg up, an unfair advantage. It's similar to the person who has read natural healing books compared to the person who just continues to go to the allopathic doctor all the time. The aware person has all sorts of advantages over the person who is just existing in the pablum of yesterday's Neanderthal models.

A whole world exists out there about which the average farmer doesn't have a clue. Your opportunity to be a farmer entrepreneur with a quality of life that would make a white-collar worker envious is directly proportional to your acquaintance with approaches that run opposite mainstream thought for the previous half century. In fact, the roots of our debacle run much farther back, but the real acceleration has occurred since World War II.

It doesn't matter what your background, your socioeconomic status, your age or your current living condition; if you have a yearning in your soul to grow things and minister healthy food to people, to live an agrarian life with your children and grandchildren playing around your feet, then an opportunity exists for you.

Write down your dreams. Write them down often. Speak of them to your relatives and friends. Seek out people who share those dreams; ignore those who do not. Don't spend time with naysayers. As your vision becomes your passion, dreams will give place to reality. My prayer is that together we can help this happen for you.

The Polyface Story

I am blessed with a legacy of farmers who thought outside the box and who lived below their means. My great grandfather, "Happy" Smith, farmed quite successfully in Indiana and had an excellent reputation in the neighborhood. His obituary mentioned his fine livestock and excellent crops. He had one son and four daughters. My paternal grandfather married one of the daughters.

Happy was out dynamiting stumps one day, clearing land, and was killed when the explosive detonated at the wrong time. My grandfather and namesake, Frederick Salatin, had married my grandmother and they tried to make a go of the farm but finally gave up. When my father was about 6 years old, they moved into town where Frederick worked as a machinist in an automobile plant. But he loved agriculture, and turned his half-acre town lot into a paradise. He grew raspberries, strawberries and vegetables. An avid apiarist, he sold honey at a local grocery store and my Dad, as a little boy, was always able to handle bees without being stung.

An octagonal chicken house in the back corner supplied eggs. He grew sugar beets and skewered them onto finishing nails about 18 inches above the floor of the chicken house to provide fresh vegetable material for the hens during the winter. Grape arbors provided the boundary all the way around the garden. Ever the creative designer, he invented and patented the very first walking garden sprinkler. Called the "Sprinkreel," it was a contraption of pipe fittings and wheels that pulled itself along by rolling up the garden hose. He was a charter subscriber to Rodale's *Organic Gardening And Farming* magazine when it came out (with the order of gardening and farming reversed). The magazine took off after Rodale put the gardening ahead of the farming.

At that time, in the late 1940s, a major separation occurred in agriculture. Many of the classics that we alternative farmers revere were hot off the press at that time. In fact, Ed Faulkner's *Plowman's Folly* was a runaway bestseller, pushing one million copies. Can you imagine such a title being a bestseller today? *An Agricultural Testament* by Sir Albert Howard was circulating along with D. Howard Doan's *Vertical Diversification.* Louis Bromfield basked in national limelight and was up for Secretary of Agriculture. Can you imagine what would have happened if he had been appointed as secretary of the USDA? He died before that happened.

At that time, the biological approach was holding toe to toe with the petrochemical approach. Of course, the chemical approach carried the day and some practitioners, my grandfather among them, sided with the organics.

I remember well the awe and reverence I felt as a little child when we would go visit Grandpa and he would take me through his garden. It was always perfect. He meticulously limited his strawberries to five daughters, carefully placing a little rock on each runner he wanted to root to weight it down right where he wanted it and keep the wind from blowing it around.

Perhaps because a couple of times we visited in the fall, I can remember his hundreds of feet of grape arbor virtually sagging with full, lush purple fruit. His compost piles, carefully stacked inside woven wire nets, always smelled sweet and refreshing. The garden had immaculate grass pathways on which the walking sprinkler ran. That meant you could walk through the garden, even if it had been freshly sprinkled, and not get muddy shoes, which must have been a godsend to Grandma.

Dad's father, then, was a master gardener, inventor, marketer and tinkerer. He died when I was about 8 years old. He had gone out to the garden and apparently felt tired because he went over under some shrubs, as was his custom, to take a little nap. He never woke up. I've often thought what a privilege that was, to pass that way.

Mom's grandfather was a master woodworker. We still have several of his woodcuts and pieces of furniture. Mom's mother lived on a city lot in Ohio and always kept a sizable vegetable and flower garden. She was a frail woman, always having one or another physical problem. From my earliest memory, she seemed bent over. But she had a strong spirit and loved to take me out into her little vegetable patch and show me what was growing. I always marveled at how clean it was because it seemed like ours at home was always weedy. She grew bushels in her little postage stamp garden. She also kept it well decorated with flowers, which was new to me because our garden at home always seemed much more utilitarian. Mom kept flowerbeds, but the garden was for edible stuff.

After Dad got out of the Navy following World War II, he went to Indiana University on the GI bill, majoring in economics. From his earliest memories he wanted to farm in a developing country. He always said he had that vision even as early as six years old. Always looking for new frontiers, he saw Latin America and South America as the place for pioneers. While in college, he gave a speech about the deplorable erosion on American farmland and endorsed conservation practices.

Although I do not know all that transpired between the two of them, I'm sure Grandpa told Dad what was going on in the big garden and what he was reading in the agriculture literature. Dad spent several months living with a family in Mexico after a stint at an intensive language school in Vermont, where he took Spanish. (He reminisced that the school had student monitors in the bathrooms to make sure that you did not speak English. When you enrolled, you took a pledge not to speak in any other language except the one you were studying while you attended.)

Then he went with Texas Oil Company (now Texaco) as a

bilingual accountant on their wildcat oil drilling ventures off the coast of Venezuela. That was a high-paying occupation at the time and he squirreled away every dime. He married Mom and they went back, eventually purchasing a 1,000-acre highland farm from an elderly señora.

He began clearing some of the jungle, built a house and built a large free-span garage. He ran a siphon from a hillside spring down to the house for gravity-pressure water and was constantly repairing it because the monkeys would gnaw holes in it. By this time, his two sons were coming on and the future looked bright. The goal was to have broilers and a dairy. Of course, a totally free market existed, without government regulations.

He took broilers down to the open-air city market and very quickly cornered the poultry market. The indigenous birds virtually all had subclinical pneumonia because of unsanitary production models. Vendors who pedaled their wares down through the city would come to the market and negotiate with the farmers. Vendors who sold chickens would run their hands down along their beaks to check for mucous drainage, an indicator of pneumonia and overall health.

They quickly learned that Dad's never had problems and were always healthy and plump. Of course, this did not endear him to the other chicken farmers, but he received top dollar. The vendors could in turn get a better price from the señoras in town when they showed up with healthier poultry. The house help would then butcher the bird on the back stoop and cook it for the family for dinner.

The tradeoff of a completely unregulated culture is political instability, and in 1959 a Communist-instigated revolution rocked the country and a period of anarchy reigned. It was the beginning of the ugly American, the capitalist pig, and we were vulnerable during the rebellion. We were not missionaries, diplomats or corporate employees. We were just Americans building our tent in a foreign culture.

People came in and took over our farm. The local constable would give us no protection without a bribe and we fled to Caracas looking for some government official who would recognize us as legitimate landholders and give us protection. People who think property rights are inappropriate should live in a culture where there is no respect for property. Essentially, the city aristocracy owns the land and lets people live on it to maintain the economy and provide cannon fodder for military campaigns.

As a result, no respect for property exists. It's just one for all and all for one, which makes long range planning nearly impossible. The surest way to destroy property is to eliminate responsibility for its care and to eliminate the profit motive for sound stewardship. Certainly pure capitalism, without morality, is not much better, as demonstrated in our own country. But capitalism with morality is positive.

Anyway, every door closed. Not a single official was sympathetic to our plight. Dad met with the muckraking journalist Drew Pearson when he visited Venezuela on one of his fact-finding missions and he promised to do what he could.

We gave away all of our possessions and sailed back to the States on a freighter. I was four and my older brother, Art, was seven. Dad and Mom were in their late 30s. I cannot imagine what an emotionally difficult time that must have been for them. They lost all their life savings and their dreams.

We were still trying to get a settlement out of the Venezuelan government for our losses. That experience forever shaped our family's view of politics and American involvement in foreign countries. Finally through the pressure from Drew Pearson to publicize our plight on his radio program, the Venezuelan ambassador in the U.S. gave us a token settlement for the farm. It was merely a token, but at least it was an admission that we had been wronged.

Meanwhile, we looked at farms within a one-day drive of Washington D.C., from Pennsylvania to northern North Carolina, and settled here in the middle of Virginia's Shenandoah Valley near Staunton, "Queen City of the Valley."

It was the most worn-out farm we looked at, but Dad and Mom thought it was the best buy, so we settled in. With that check from Venezuela Dad bought 28 Hereford cows and started farming again. This farm had been part of a much bigger farm in the late 1700s and eventually passed out of that family around 1890. It was divided five ways, with each of the children getting one piece.

Our farm, which consists of two of the pieces, passed out of the family in about 1915 when it was sold at a sheriff's auction on the courthouse steps. It was absentee-owned for several decades thereafter, being rented to neighbors. Being in the grain-growing region of the mid-Atlantic area, the land had been plowed for small grain for a century. Since horses did not turn over on steep hillsides, everything was plowed and huge ditches began. Up to eight feet of topsoil had eroded away from some hillsides. Most of the steep hillsides had gullies like corrugated roofing running up and down them when we came in June of 1961.

Dad immediately focused his attention on conservation efforts. We planted many acres of trees until the original 160 open acres shrunk to only about 95. Some areas were so rocky we couldn't plant trees. We fenced out these areas and let them vegetate naturally into mixed hardwoods and evergreens. He sought counsel from many sources, both public and private, for how to make the farm pay. I can remember consultants coming and walking over the farm, talking with Dad and Mom late into the night. Their advice was consistent: "Graze the woodland, plow and plant corn, make silage, and apply chemical fertilizer."

Dad eschewed all that advice. He read some André Voisin material and realized that the secret to productivity was in managed grazing. He invented a portable electric fencing system and we began moving cows around about every couple of weeks. At that time, fence chargers used points and condensers from cars to break the spark. The points were forever corroding.

After only one year he realized that the farm could pay a mortgage or salary but not both. He immediately liquidated the cow herd and took an accounting job in town. Mom began teaching health and physical education at the local high school, and they set about to pay off the mortgage. In just eight years we were debt-free.

Dad turned his attention to research and development. Since our milk cow gave more milk than our family could consume, we began buying baby calves at the stockyard and raising them. Dad built a portable veal calf trailer, 15 feet x 15 feet, with a slatted floor. We bought a couple more cows, divided the veal trailer into four quadrants, and let the calves hop out and nurse their cow at milking time. The cows shaded up under the eaves of the veal trailer and we moved it around the fields to spread the calf droppings. It worked fairly well but Dad's off-farm accounting work, which required late nights from January 1 to April 15, and occasional late nights auditing for certain clients, kept him from being able to keep a good schedule. Milk cows and baby calves need a consistent schedule. He finally shut it down, but the concept works.

During this time, the fertility was so low that we scarcely had enough hay for 15 cows. We made loose hay with an old hay loader and Dad developed a feeder gate that would walk through the haystack to make feeding easy.

My older brother, Art, raised rabbits and I distinctly remember driving one Sunday afternoon to visit a man who had portable A-frame pig huts out on pasture. Dad used that idea to build some rabbit pens but they dug out. He and Art built some rectangular pens that fit up to a four-quadrant rabbit house. But the rabbits dug out of those too. We put finishing nails down from the bottom and clipped off the ends to make them sharp, but that didn't help. We put a foot-wide wire fringe around the inside perimeter and that helped, but reduced the amount of grass the rabbits would eat by half, which seemed to be a waste of space and time.

A poultry netting floor kept the rabbits from digging out but also bent over the grass when we moved the pen. The rabbits would have to reach through the netting holes to graze and would mistakenly bite down on the wire netting. That made them reluctant to graze. Furthermore, with a netting floor, we had to pick up the weight of all the rabbits when we moved the pen and that made it too cumbersome.

Finally we abandoned the pasture rabbit pen idea. About

that time I was 10 years old and wanted an enterprise so I settled on laying chickens. My favorite great uncle was quite a chicken farmer in Ohio and I always enjoyed his chickens. I began with a little flock and it eventually grew to a hundred.

We had these rabbit pens sitting around and Dad suggested that I put the chickens in those and see how the chickens would do being moved around on pasture. Little did we know that we were seeing the birth of the pastured poultry model.

By 1970 I was 13 and producing quite a few eggs. Early on Dad realized the futility of trying to make money on a small farm by selling wholesale. As an economist, he saw the potential of selling everything retail. I joined 4-H in order to sell at the local Curb Market, a precursor to farmers' markets. This market was left over from the Depression era, when farmers were notoriously short on cash but long on food. It was a mechanism for giving farmers an outlet for their food, and city folks a chance to purchase what was being grown in the countryside. The market grew to a place of real prominence in the community during the 1940s and finally began to ebb in the 1950s.

Most of the vendors were women, and they joined the Home Demonstration Club in order to sell. Today, these are Extension Homemakers' Clubs. The market enjoyed exemption to nearly all inspection regulations on meat and dairy products. By 1970, it had dwindled to just two elderly ladies.

One came from a diversified farm producing vegetables and pork. She also made pastries. The other lady lived in town and sold baked goods, pastries and potato salad. I owe my love of potato salad with sweet pickles to her. If she didn't have potato salad, I would get a wedge of pound cake and wash it down with my raw skimmed Guernsey milk. To this day, pound cake and sweet pickle potato salad are two of my most favorite foods.

We opened a booth there and I sold butter, buttermilk, cottage cheese, yogurt, fresh beef, pork, rabbit, chicken, eggs and vegetables. One of my big problems was what to do with spent laying hens. The going rate at the time was 19 cents per bird. I would come home from school and get my slave, younger sister Loretta, to help me heat a metal pot on a wood fire and dress 20 or 30 old hens. Then during the week I would roast half a dozen at a time in Mom's big roaster pan and pick the meat off while listening to Paul Harvey's record *The Uncommon Man.* We never had a TV (and still don't). Then I would freeze that chicken in quart containers and sell it as picked-off, precooked chicken at the market for $1.50 per pound. My Leghorn hens would give a pound of meat per carcass, which meant I was getting nearly 800 percent more than the going rate for spent hens. All through my high school years while classmates were sleeping in on Saturday morning from their foolishness on Friday night, I would rise at 4 a.m. and head down to the market, which opened at 6 a.m. By 8 a.m. I would sell 60 dozen eggs. I had both white and brown varieties.

I would not trade that experience for anything. Here I was, a teen, with these two elderly ladies who had more old-timey stories than a roomful of men put together. They told how "the old people" did it; they showed me how to display my produce, how to handle complaining customers and how to turn a fence sitter into a sale.

We had slow mornings and bustling mornings. In those days, nobody knew what "organic" meant. They just wanted local and fresh. They knew what farm-fresh food tastes like; they were not concerned about food safety and chemical residues. Our clientele steadily grew and Dad saw the potential of moving a beef per month down there at retail prices. As stockyard prices rose and fell by 100 percent or more, our prices held steady right with retail prices and the customers kept coming for this grass-fed beef.

Finally I graduated from high school and we closed down the stand, probably putting the final nail in the market coffin. By the time I returned from college, it was closed up . . . forever. One of the ladies had passed away and the other was not doing much anymore. I often wonder if this energetic young man didn't heave a last gasp of breath into the market for these two older ladies.

Although I do not remember the conversation specifically, Dad often reminded me of it: the evening I told him I wanted to come home and farm. He said we were standing "at the dogleg," a bend in the road behind the house. A realtor tax client had just told him what the farm was worth. Ever ready to embark on a new venture, he realized that we could sell this place and buy one twice as big in a more deflated area.

It was in that context that we conversed about the future. I was a junior or senior in high school, but I was convinced that this was where my heart lay. The farm had not really made any money, but we had had some successes. Fertility was gradually increasing and my marketing ability was obvious. Dad had the economic and engineering skills; I had the people skills. Mom coached the debate team and that is no doubt where my communication skills developed. We still could not see how the farm could actually pay a salary. So I decided to go to college and become a journalist, perhaps write a best seller, and after a period of years retire to the farm. After all, you had to be independently wealthy to farm.

I worked at the local newspaper during my junior and senior years in high school and got along well. When I got out of college, I was a shoo-in for investigative reporter because all the staff already knew me. During the college summers, I worked on the farm. When I graduated, I had several months before I started my off-farm job. I spent that summer of 1979 installing our permanent electric fence system. That kept us from having to move so much fence every time we changed paddocks. It also meant that our paddock stays shortened from a couple of weeks to a few days.

To illustrate the genetic base of my deep passion for actively pursuing what is right, during the mid 1970s when the Arab oil embargo was in full swing, Dad began riding a bicycle to his off-farm job 14 miles away. He was in his 50s at the time, and decided this was something he could do to break the yoke of petroleum. That is conviction.

By this time, my future wife, Teresa, and I had begun making definite plans for living on the farm. We brainstormed about being able to hand milk 10 cows, sell the milk at retail prices, and make a comfortable living. I well remember the figures in that exercise, because at that point I began realizing that maybe we could make a living on the farm. Dad and I were devouring *Mother Earth News* and I saw how cheaply we could live if we put our mind to it. I began writing for the newspaper and living at home. Finally I bought a car for $50 from a friend.

By living at home and saving, saving, saving, I was able to get a little nest egg started. Teresa and I married and took our honeymoon in that $50 car, which I sold the next year for $75. We've now been married 17 years and have not yet spent $6,000 total on automobiles. We fixed up the attic in the farmhouse (we called it our penthouse). Our bedroom was also the kitchen and we pinched pennies hard enough to make Lincoln scream.

As soon as I came back from college, I began raising baby calves on our extra milk from the family cow. I sold one calf the following year to a friend in our church group and a guy at work — a half to each. They both loved it and I got 40 percent more than I would have gotten at the stockyard. We were onto something.

Teresa and I lived on a couple hundred dollars a month. If we didn't grow it we didn't eat it. We never went out for entertainment. The farm was our life, our recreation and our love. She made all our own bread, sewed our own clothes, and put a heart into our home.

Meanwhile, I was doing more and more agriculture writing for the newspaper since I was the only one on staff that had interest or understanding in that area. That put me in touch with the agriculture community and with farmers. I heard horror story after horror story and realized that our farm had to be completely different from what I was seeing if we were going to make a living at it. I saw sick animals, bankrupt middle-aged farmers. Countless young guys my age who had inherited farms lost them amid all the emotional and mental anguish that can occur in a multi-generational business.

These were precious lessons, seeing and learning what does not work. I saw behind the curtain of the charade and realized that for all the promises and posturing of the agri-industrial complex, it is a dead-end street. Seeing the unhappy farmers, sick animals and collapsing bank accounts juxtaposed with the smug public relations personnel at press conferences simply strengthened my resolve to do things differently. For all their platitudes about feeding the world, reducing disease and maintaining the U.S. economy with agricultural exports, the agri-industrial complex was rotten to the core. I saw the dead animals piled up outside the concentration camp factory houses, listened to countless farmer horror stories, smelled the drugs and manure. All of this strengthened my resolve to have nothing to do with the agri-industrial monster and to successfully chart a different course.

I sat through hours and hours of USDA informational meetings, attended seminars and field days sponsored by government agencies and corporate patrons. The average person cannot imagine what farmers are routinely told. Certainly I found nuggets of gold in some of these programs, but 95 percent of the presentations steered folks in the wrong direction. And then when I visited farms and saw the result of conventional recommendations, I saw just how evil the advice was. Although I didn't know all the answers, I knew where *not* to find answers. And during this time, our farm, embryonic though it was, showed real promise. We had some real successes in soil fertility, healthy animals and low-capital production models. The contrast between what I was doing at home and what I was seeing in the countryside was dramatic indeed and filled me with hope for our farming future.

A black walnut buying company, Hammons Products Company in Stockton, MO, began buying walnuts in Virginia in 1981. Two local FFA high schoolers ran the hulling machine and I went down to do a story on this new enterprise. I saw pickup trucks loaded with walnuts lined up for half a mile while these two boys, their fathers and some friends worked from before daylight until after dark every Saturday morning from early October through November, handling the volume of walnuts.

Their biggest problem was what to do with the tons of hulls generated in the hulling process. From day one, we have been fiends for organic material. In the early 1960s Dad would drive our 1952

International dump truck to town and bring home truckloads of any organic material he could find: dried sludge from the sewage treatment plant; corncobs from the grain elevator; sod from the road side ditch cleaning operations. So when I saw this mountain of hulls, I volunteered to send Dad in the next Saturday for a load.

The boys were grateful, and we learned how the process worked, what the boys were getting paid, and how beneficial the hulls were once we spread them on the land. Grass grew like crazy.

I learned that the local Southern States store, where the huller was set up, was snarled because of the congestion caused by the traffic line waiting to get to the huller. The administrator decided that if they were going to let the huller be there again, it would have to be open six days a week, during regular store hours. From working on the news angle I knew about the need so Dad and I pushed the pencil on it. We decided to make an offer to the store to run the huller the following year.

The store was delighted to have someone willing to take the responsibility for the walnut buying station. Teresa and I decided that we had enough money saved up that with the commission from the black walnut station we could survive for one year. Our hope was that we could survive longer than that. But even if we did not, we figured that the year at home would get us so much closer to our goal of full-time farming by enabling us to accomplish a number of tasks that perhaps it would shorten my total time away from perhaps ten years down to five.

I gave my two-week notice and walked out of the newspaper office, to the incredulity of everyone ("You are crazy! Don't you know you can't make any money farming?") on September 24, 1982. A week later, I was running the walnut buying station and bringing home truckloads of black walnut hulls.

Unfortunately, that year the crop was only a fraction of what it had been the first year. We have since learned that a bumper crop only occurs about once every five years. We tightened our belts a little more and resolved, by the grace of God, to stay with it, to persevere as long as possible.

What happened is an important lesson to remember. Although we made almost no cash, our living expenses dropped drastically. Instead of filling up the car with gas once a week, we filled it up about once every six weeks. I didn't have to buy any dress clothes. My good shoes did not wear out. I was home for every weed that grew in the garden, each calf birthing and each perfect compost-spreading day. By reducing both expenses and what I call on-farm slippage, we eked through the first year.

Then the second year some Amish friends asked us if we'd like to take over their little backyard broiler operation. They grew about 400 a year and tried to make enough that they got their own personal birds free. We had not had chickens for about six years.

We decided to go for it, so we broke out the old metal pot and an ancient feather picker and got into chickens. We eased the old chicken pens down from the rafters in the barn, where we'd pulled them up for storage after I got out of laying hens when I went to college. The broilers did great and we were off and running with a new enterprise. We sold two beeves instead of just one and added some firewood sales. We were still selling most of the cattle at the stockyard.

A year or so later Dad retired and then Mom retired. Teresa and I began assuming more and more of the farm expenses, including insurance and taxes. From the day he knew I wanted to farm, Dad quit buying machinery, preferring instead to have me own it all. That way I would take better care of it and it would not accrue to the estate.

Dad passed away in 1988 but he knew we were onto something big. We were selling everything retail, customers were coming in droves, and we were debt-free. He would say: "Son, you're on the right track. Just stay with it." And we have.

With his and Mom's launching pad, Teresa and I, and now our children Daniel and Rachel, are blessed beyond measure. We are enjoying the fruit of Mom and Dad living a little below their means. Usually, each generation lives just a little above their means.

This tends toward impoverishment. Look at consumer debt in this country and you will see what I'm talking about.

If each generation will live a little below its means, true wealth can accrue. It is our prayer that our children and their children will never forget Mom and Dad scrimping and scrimping, making Christmas gifts instead of buying them; refusing to have TV or participate in all the busy activities children get into; turning their backs on all the advice to plow, chemicalize and increase their debt. We trust that our children will never forget that neither Teresa nor I owned a car until after college; that we turned our backs on lucrative "city work" and turned our eyes toward home and farm; that we patiently plodded toward a dream, and disciplined our short-term gratification in order to enjoy long-term success.

All the while we've been pursuing this path, friends and family have told us we are crazy. Other farmers think we're lunatics for farming the way we do. Friends who mean well expressed heartfelt concern about throwing away our lives. College friends wondered why we would squander our education to do something as menial as farming. While our friends were building houses, accumulating mortgages and buying shiny cars we wore ragged blue jeans, shopped at the thrift store and drove clunkers. In fact, we've never yet owned a car in the same decade in which it was built.

I want to know where the people are who will set their eyes on a goal and go for it in the face of opposition. Where are the folks who will dedicate themselves to pursuing a dream, regardless of what others will say? Where are the young people who will limp along on one salary, drive one automobile, live in a tent or used mobile home in order to stay financially solvent and flexible?

Where are the parents who will toss out the TV and homeschool their children, or encourage them in debate, drama and forensics instead of jockey sports? Learning to communicate will be far more valuable than learning how to put a ball through the hoop. I hear people say that I'm so unusual I would have been successful at anything, so I have no credibility to tell people how to farm. Are you willing to do what Teresa and I have done?

I'm not trying to be arrogant; I'm just trying to make the point that if we are not becoming what we want to be, we must make whatever changes in our thinking and our lifestyles necessary to become who we want to be. It's time to stop making excuses. It's time to stop listening to naysayers and head-waggers.

It's time to quit living on the fast track to stupidity and get on the right track to fulfillment and success. It can happen; it does happen. But successful people are not born; they're made. I cherish my background of working on the farm, having my own business beginning at the age of 10, participating in drama and debate, having no TV and caroling with the church group at Christmas. This is the stuff of character, of passion, of dreams realized.

The last several years have been nothing short of miraculous. Our models work; our hard work is paying off. We are now embarking on a new frontier — how to touch people with the vision we have and the information we've gained without jeopardizing the economic viability and productivity of the farm. This next era will be just as challenging as building the models in the first place. Try bringing in a couple of apprentices and you'll see what I'm talking about.

But life should always be a challenge. We press on, ever seeing our vision widen: ***to develop emotionally, economically and environmentally enchancing agricultural enterprises, and facilitate their duplication throughout the world.***

The Right Philosophy

Our philosophy underpins everything we do. It determines where we go to find information, how we assimilate that information, and how we make decisions based on that information. Having the right philosophy is unquestionably the starting point for any successful venture.

Your philosophy and mine may disagree, and that is okay. But to be fair to you, and to be completely open about where I'm coming from, let me explain my philosophy regarding the farm, biological communities, and our culture. All of my advice stems from this mindset. Even if we are not in complete agreement, I trust you will find more with which to agree than to disagree. If we disagree on almost everything, give yourself a big pat on the back for at least looking at an alternative.

Every one of the world's great religions, from Judaism to Christianity to Hinduism to Muhammadanism includes "sacred" writings denoting a reverence for nature. The "fearfully and wonderfully made" elements of the *Bible's* Psalm 139 bespeaks how awesome and incomprehensible creation is.

Scientists still do not know what holds atoms together. Yet in the *New Testament*, Col. 1:17 says: "And he is before all things, and by him all things consist." The "life force" described in Oriental literature, and especially in the writings of Rudolph Steiner, founder of biodynamics, give weight to the metaphysical, or the importance of life aspects beyond mechanics.

Until just the last two centuries people held a reverential respect for life, for creation, and for the complexity of nature. Very primitive tribes around the world possess a deep reverence for a Creator, and the inexplicable handiwork of the universe. The sacred land element within the breast of many native Americans underscores a reverential attitude toward nature.

While I do not condone animism and the worship of nature, and certainly appreciate that reverence too often gives way to worship, by the same token I do not agree with the notion that life is just so much protoplasm to manipulate and adjust however we see fit. The philosophical underpinnings that gave us DDT and are now giving us genetic engineering would have been completely foreign to any of the world's religions as recently as three centuries ago.

How have we gotten to this point? Justus von Liebig (1803-1873) is known as the father of modern chemical agriculture. He isolated nitrogen, potassium, and phosphorus (NPK) as prime elements in plant tissue and deduced that if we would but replace those things in the soil, we would have the building blocks of plants.

It was a very simplistic approach, but a mechanistic approach. He was not concerned about "holism," a term coined by Jan Smuts in South Africa when he realized nothing could be studied in isolation. No matter how small the unit, it was still part of a whole, and to isolate it from that whole removed it from its world of interactions and it ceased to be exactly what it was.

Liebig made no differentiation between synthetic and natural, and had no concept of all the microorganic community in the soil that brings nutrients to plant roots. He did not study the nutritional aspects of his NPK plants, or the health of the people who ate them. It was as isolated, as "unwhole" a research system as you can imagine. Part of what made Liebig's discoveries so damaging was that after he reached middle age, he became a prolific writer. This spread his thinking to a much broader audience than if he had stayed with his vacuum tubes and laboratory.

While Liebig was doing his work in Germany, two contemporaries were working in other parts of the world. An Englishman named Charles Darwin (1809-1882) was appointed the naturalist on the touring ship, *Beagle*, which sailed from England in 1831. Of course, we all know how he took God out of creation and propounded evolution. It is significant that even today, when primitive tribes around the world first hear about the notion of evolution, they react with hilarity and incredulity. They can't imagine anything so absurd.

Across the Atlantic, the very year the *Beagle* sailed from England, an American inventor-blacksmith in Steeles Tavern, Virginia demonstrated his reaper. The reaper injected machines into agriculture on such a dramatic scale it completely altered farm life from being one primarily concerned about living things to being one equally concerned about the machinery.

John Deere developed a steel moldboard plow and the combination ripped open America's prairie, converting it to grain fields. While I do not have a problem with technology and machines per se, I think it is important, if we are to keep any moorings to the truth, to appreciate how the simultaneous events described here entered the thinking, the philosophy, of the world, and especially the Western mind. The Western mind, devoted to logic and reason rather than eastern mysticism, quickly accepted these mechanical changes because linear, reductionist thinking did not ask bigger questions about how these things impacted ecology.

Indeed, the word "ecology" was not invented until 1869, coming from the Greek root *oikos*, meaning "household." It was literally the "housekeeping of nature" and grew out of a need to develop a non-mechanistic mindset toward living things.

Do not miss the compound effect that all of these men, operating in different parts of the world but at the same time, had on the world's thinking. While Liebig told us plants were just elements stacked up in a certain way, Darwin reduced creation to happenstance and therefore life to amoral chance, while McCormick and Deere revolutionized food production away from being primarily a biological process to primarily a mechanical process.

With all this in mind, imagine the impact of Henry Ford (1863-1947), whose horseless carriage chugging down the street of Detroit in 1896 paved the way for the modern factory. He founded the Ford Motor Co. in 1903 with one primary goal: mass production. That's what the encyclopedia says.

Because of this goal, he developed the interchangeable part and the assembly line. This paved the way for millions of people to be factory workers instead of self-employed craftsmen and small businessmen. With the Creator out of the picture and life reduced to so much inert material, America's landscape was ripe for exploitation rather than husbandry, rape rather than resource stewardship, and adulteration rather than reverential awe.

As Western culture projected this mechanistic, all-controlling, manipulative philosophy onto nature, and onto agriculture, true life energy, food nutrition and purity began leaving America's food supply. We poisoned our water, destroyed soil organic matter, eroded the hillsides, choked fish and did it on such a grand, glorious scale that now health care is the nation's single largest business. Why? Could it have anything to do with how we've stewarded what God wanted us to adorn and nurture?

By no means am I a tree-hugging, animal-worshipping environmentalist; but neither do I agree with mainline conservative thought that the most glorious sounds are of chainsaws and vaccination syringes. Nature is balanced between floods and babbling brooks, between blue skies and thunderstorms, between hot and cold, wet and dry. To say that a cow is sacred is just as unbalanced as to say a cow is just a big pile of protons, neutrons and electrons. To say that a farm is a factory is just as wrong as mysticizing a farm to the point that we say it is not even a business. To say that no cow should ever set foot in a creek is just as wrong as saying it doesn't make any difference if the cows live in the creek. We must be balanced.

With all this in mind, let me articulate some of my broad philosophical concepts, specifically as they relate to the farm.

Environmentally enhancing agriculture. There is simply no excuse for any type of agriculture that degrades the environment. I am not a believer in "trade-off" mentality. I do not believe for an instant that in order to produce enough food we need to sacrifice environmental quality. Included in this goal is smell.

Any food production system that stinks up the neighborhood — regardless of how rural — is unacceptable. Excusing farm smells with that euphemistic "fresh country air" business is ridiculous. If you ever smell manure, you're smelling mismanagement.

What we want are farming models that actually leave the environment better than it was before we came. We want the soil more fertile, the landscape more diverse, the forest healthier and the wildlife more prolific. And yes, I do believe people can improve on "pristine" nature. For example, one of the best ways to stimulate biodiversity in a wetland is to build a pond. This would make a government-has-all-the-answers environmentalist shudder, I know, but the injection of the human mind and labor into a landscape can make it far better than it was as long as the changes are within the parameters of proper environmental considerations: clean air, clean water, healthy plants and animals.

We ran into this when we began marketing eggs into upscale Washington D.C. restaurants. The chefs wondered why our eggs were variable: spotted shells, dark brown, light brown, some more pointed, some more round. They expected these eggs to be cookie-cutter perfect. I explained that the variability was because we were using old-fashioned non-hybrid laying hens and they exhibit much more genetic diversity than hothouse, hybrid chickens.

This just goes to show how far away from reality people can get. If the height of environmental activism is to send a check to the Sierra Club, as important as that may be, one is not much of an environmentalist. Until our thinking translates into life-style action, we're just playing games with our convictions. This brings us to our second goal.

Bioregional food sufficiency. The average T-bone steak in America has been in more states than the farmer that grew the beef. Keeping our production, processing and marketing within the same locale stimulates the economy by keeping all the value-added dollars home.

Advocates of the macrobiotic diet say that food produced more than 40 miles from your home contains a different type of energy. I don't know about that, but it only stands to reason that in nature, things eat local food. Predators do not ship in prey from 1,000 miles away. Soil bacteria ingest organic matter from decomposing plants that grew in the immediate area. Furthermore, the community becomes less dependent on transportation when folks feed themselves. Right now, it takes 15 calories of energy to put 1 calorie of food on the American table; 4 calories of that are for transportation.

Of course, those who disagree always ask: "What about New York City?" I used to be somewhat apologetic about not having an answer for that, because clearly the huge city is too large for the nearby countryside to supply all its food. Importing from distant areas is essential to feed a large city. But I have changed my response to a question (Socrates would be proud): "Why have a New York City?"

I'm not being silly. I view huge cities as completely unnatural. They do not produce food, water, air. They have sewage problems, solid waste problems, pollution problems. Why should anyone be interested in preserving an inherently inefficient human living model? Let's agree that huge cities need not be preserved in their present form, and each go about promoting in our own locale a more sensible model of existence.

Wendell Berry, one of my favorite writers, says that there are no global problems; only local ones. Most of us spend a lot of time and money dealing with and worrying about things that we can't do anything about anyway. If we would devote that same energy to our little realm of influence, the cumulative effect would be a much better society.

Seasonal production cycles. This goes hand in glove with bioregional food sufficiency, because it speaks to disciplining our desires to be compatible with nature. As soon as we require unseasonable production, the farmers' costs go up and the caloric production requirements increase.

Why should Virginians in January import fresh tomatoes from the salinized San Joaquin Valley? The extra demand for the vegetables puts a strain on those small areas that can produce the food and the transportation requires huge amounts of petroleum to get it here. Why not freeze, can or dry the extra truckloads of tomatoes Virginia produces in the summer — at much lower prices — and eat them during the winter?

Beef is a heating meat; chicken is a cooling meat. Beef should be processed in the fall, right before winter, when it is of highest quality. Then it can be consumed during the winter when we need the heat.

Chickens lay extra eggs in the early spring — more eggs than we can eat. The longer daylength stimulates production. The extra eggs can be hatched for broiler production. Broilers grow out to dressing weight in just 6-8 weeks, right on time for those hot summer days when we want a cooling meat. The extra broiler chickens can eat the additional bugs that come in the summer.

Vegetables cool us down, and grow in the summer. Through careful planning, fresh fruit can be harvested throughout the season: strawberries first, then raspberries, then blackberries, then grapes, then apples and pears. Fish is a cooling meat if it comes from warm waters; fish from cold waters have more calories. The more we view our natural world through the eyes of wonder and discovery instead of dominance and arrogance, the more reverential we become.

Bucking the system by starting baby chicks in the middle of winter, calving in the middle of winter, all take a toll on efficiency, fun and economy. Following the seasons breaks up farm monotony so that we have busy times and restful times; animal times and construction times; people times and solitary times.

Decentralized food systems. We need an agriculture that stimulates "spread out" modeling rather than centralized production and processing. Currently four beef packers control 80 percent of the U.S. market. In the grain belt communities that used to have several competing grain elevators now often have only one, owned by one of the huge multinational corporations. The "grain cartel" is a household word in agriculture, and is merely a manifestation that we do not have the competition we once did.

The poultry and swine industries, likewise, have become highly centralized. Unfortunately, the grain and the animals are not in the same area. The grain is exported from regions that desperately need the animal manures produced through consumption. The animals are concentrated into areas that can't utilize all the manure. This concentration denies one area its fertility and foists upon the other too much manure. Groundwater contamination, smelly production systems and unhealthy animals result.

Food-borne pathogens get mixed and matched in huge processing facilities to infect hundreds of people in remote areas. When will people understand that putting a million birds through a processing facility is not the same as putting a million copper fittings through a factory press? Rather than stimulating smaller facilities, backyard operations and neighborhood canneries and processing facilities, consumer advocates demand salvation by regulations which destroy additional small operators.

Of course, centralization, or mass production, as Henry Ford called it, reduces genetic diversity because assembly lines must run within close tolerances. It just won't do to have a carcass that's 10 percent bigger or smaller, or a tomato that's a different size. And especially it won't do to have an apple or tomato that is so tender it can't withstand mechanical picking, sorting and transportation. All the considerations that anyone in their right mind understands to be important — germplasm diversity, nutritional quality, taste — get thrown out in the quest for standardization.

Then folks at Archer Daniels Midland use synthetic this and that to mix and match and make these junk foods edible and create concoctions the human body was never designed to ingest. And all this is supposed to help the farmer, who has been reduced to serfdom, cranking out the widgets the industry demands. No room for developing animals that perform without pharmaceuticals; no room for selecting for an apple that has higher vitamin C or Brix than others.

If we are to see this stranglehold on America's food supply slackened, we must reduce regulation, increase freedom, and unleash the entrepreneurial spirit that is alive and well in America's countryside.

Entrepreneurial private sector small business. I was pleased to learn that just recently mainline environmental groups have begun to see small farms as part and parcel of the solution to the macro-problems these groups cuss and discuss. Generally, these earth-friendly groups who demand salvation by legislation have enjoyed large farms because the fewer there are, the easier it is to regulate them.

The huge environmental and social degradations so commonly documented in the media have not come from cottage industries and small farmers — they have come from arrogant, sophisticated centralized well-heeled industrial complexes. Entrepreneurs are not the enemies of truth. The enemies of truth are the folks who, because of their improper production, processing or marketing model, must prop it up with misrepresentations and innuendo. When the chemical companies say that these pesticides or herbicides "break down in the soil," it is a half-truth. They do break down in soil extremely high in both living and dead organic matter. But most of our soils, through the use of chemical fertilizers, have become so low in organic matter that there is not enough left to break these compounds down.

The chemical companies say we're using fewer chemicals. Yes, but the ones we are using are more potent, so that an ounce now is equivalent to a pound a decade ago. And we've traded DDT for genetic engineering.

Entrepreneurs, because they are on the front part of the business curve, have far more to lose with slipshod practices and playing fast and loose with the truth. To preserve these folks in agriculture requires breaking down barriers to markets, opening up new choices to consumers and unleashing cottage industry creativity.

I cannot sell a glass of raw milk to my neighbor. In our community folks used to get together for a hog killin' at Thanksgiving and at the end of the day the neighbors would buy some sausage, a ham, and take home a pan of fresh ponhoss. Not today.

My daughter, Rachel, can bake the best pound cake in the world but can't sell it to her grandmother because it wasn't done in a government-inspected kitchen.

I did a marketing presentation in Ohio a few years ago and shared the symposium with a bureaucrat who began telling the folks about making windshield washing fluid out of corn. Nobody can do that without a million-dollar factory in their backyard.

When it was my turn, I asked the audience: "If you could milk 10 cows by hand and sell the milk to your neighbors, making $25,000 a year on your 25-acre farmstead, how many of you would do it?" Nearly every hand went up. That is the spirit that is alive and well in America, but through government regulations we have denied these opportunities to farmers, and denied freedom of choice to consumers.

If present trends continue, within another generation the only thing available to the American consumer will be BST milk, genetically engineered plants and animals, and fabricated concoctions out of the Archer Daniels Midland refinery. Don't get me wrong — I am not interested in regulating these folks either. Let them go about their stupidity.

But if we entrepreneurs can "have at 'em" in the marketplace, and present food that's really nutritional, that really satisfies, that really tastes good, that really keeps them out of the doctor's office and gives them roots to the land, we'll smoke those big boys. Won't it be fun?

Humane animal husbandry. Now I'll bet you conservatives who "amen-ed" that previous section are having a hissy fit.

Don't. Again, I'm not an animal worshiper. But I do believe animals should be given a chance to express their animal-ness. A chicken should be allowed to be a chicken; a cow a cow; a pig a pig. Putting a pig on slatted floors in little cages over a slurry pit does not allow the animal to express any of its instinctual "pigness."

I once walked into an egg layer house. Here were three tiers of cages as far as the eye could see. The cages were 14 inches by 22 inches. Inside were 8 chickens. They didn't even have enough room to sit down. They just milled around the cages all the time. No nest. They just squatted and dropped the egg on the slanted wire mesh cage bottom and the egg rolled down onto a conveyor belt. The birds were debeaked so they wouldn't cannibalize and most cages had at least one dead bird in some stage of decay. Finally, after the other birds walked over the carcass long enough, the decomposed carcass fell through the wire floor and went out with the manure.

These chickens are confined for roughly 10 months like this, until they burn out and a new set is brought in. Folks, you don't have to be an animal worshiper to find that treatment reprehensible, abhorrent, and any other adjective that's printable. It is indescribably terrible and ought not to be. Is the answer regulation? No. The answer is for people to quit buying those eggs, to vote with the pocketbook, to patronize their local producers.

The answer is also for you as a producer, as an up-and-coming farmer, to promise yourself that your animals will be able to express their instinctual desires. Commit yourself to happy, stress-free animals.

Relationships between rural and urban areas. Modern American agriculture has done everything possible to build barriers between city folks and country folks. Farmers erect "No Trespassing" signs because their animals are so fragile they are paranoid of disease. Country folks think city folks are ignoramuses and city folks make wise cracks about "rednecks" and "country hicks."

City folks think country folks are out to destroy the environment. Country folks think city folks don't have a clue about how the food gets on their plate, and that these city folks are trying to destroy farmers, their property rights and freedoms.

Both sides mistrust each other in the worst way. But neither will make a move to reach out to the other. About the only thing farmers do is "Ag in the Classroom," in which city kids see factory confinement houses and are told that smells just come with the territory. City folks, on the other hand, encourage regulations to stop the abuses they see on *60 Minutes,* and the result is that both sides move farther and farther into misunderstanding, animosity and mistrust.

Because farmers have voluntarily withdrawn from the marketplace — remember when farmers peddled hams, eggs and fresh chickens down Main Street? — we have created an agriculturally illiterate consuming populace. I've even met gourmet chefs who thought you had to have a rooster with hens in order to get eggs.

Consumers who care about their food need to understand that security does not come from bureaucrats or labels; the best way to know the quality and safety of your food is to see the source and visit the farmer. By the same token, the farmer who wants to educate folks about his needs should begin building relationships with them so they won't regulate him out of business.

And as we begin building relationships, we can solve problems instead of creating them and build bridges instead of barriers. Such a change will not come from labels, billboards or slick advertisements. It all starts with a relationship.

Rural non-industrial development. The Shenandoah Valley, where I live, has been quite rural. But it seems like every time I pick up a newspaper and see what's happening in local politics, some bigwheel is saying that the answer for economic development in the county is industrial expansion.

The industrial expansion brings more house-and-lot folks, which requires more schools, police and utilities. The tax burden goes up and agriculture picks up the tab. If what the conventional mindset calls "growth" were really the answer, New York City would have the healthiest economy and the healthiest government budget anywhere. But it's been bailed out by the state. Clearly, what the economic development gurus think is helpful is actually devastating.

Right now, Virginia loses about $1 million per day in balance-of-trade deficit on beef. Just on beef. We export stocker calves at $300 apiece and import them a few months later as frozen, boxed beef, for $1,200. Along the way, of course, they were transported, fattened, processed, and imported. All that added value.

What this amounts to is a state shortfall of $365 million per year just on beef. What would it do to our rural economies if we could have all that money? That's a lot of money for a little state. The same could be done for poultry in Illinois or sweet potatoes in Oklahoma.

Local canneries, little sawmills and backyard butchers built thriving, bustling rural townships and villages. It will take the same infrastructure to bring them back. Plenty of jobs exist in the countryside doing things that farmers and local craftsmen have voluntarily given over to huge factories and Madison Avenue ad agencies. A recent survey of industrial swine operations in North Carolina showed that each facility created 20 jobs but displaced 100 farm families.

Rural economies are reeling under the pressure of industrial development authorities which wine and dine, give free roads, give tax-free bonds and special tax concessions to get "more jobs." Folks, "more jobs" are right under your noses if you'll let local farmers produce and process your food, let local cottage industries sew your clothes and let local canneries use summer excesses.

Biodiversity and soil building. We're in the business of producing soil and diversifying the landscape. That means we want to see the soil more fertile every year. It should hold more water, produce better plants and require fewer inputs.

We want to see more wildlife, more plants, and produce a wider variety of animals. We want more intersecting ecosystems of water, land and forest. That means we'll cut some trees to stimulate brambles and saplings. We'll not touch some areas to keep mature trees, and we'll develop some fields into the forests to create more edge.

We'll scoop out some ponds to hold water runoff and trap

sediment. We'll fence out riparian areas to stimulate elderberries, cattails and deer tongue. We'll fence out woods to stimulate regrowth and leaf litter. We will not use chemical fertilizers that kill soil life; we will stimulate field mice and chipmunks to feed foxes and raccoons so they will not prey on our chickens.

We will use our animals in symbiosis, so that each complements the other and the whole is worth more than the sum of the parts. As we begin closing our cycles, we can produce more animals which produce more manure which we can compost and build more soil. Perhaps we'll have to replace the fences because the soil will build up over the fence.

Family-friendly agriculture. Our models must allow us to incorporate the whole family into the enterprise for mutual benefit. When a farm is nothing but confinement houses, motors, PTO shafts, silos (bankruptcy tubes), concrete and machinery, there aren't very many things for children to do.

But on our farm, children can go with us on every job. Even when we cut trees in the woods, a child can go because we're not using big machinery. We need not worry about them falling into a manure lagoon — instead, they can eat the compost and it won't hurt them. We need not worry about them wandering into a poison room because we don't have one.

When Rachel was a toddler, she would go out to the chickens with me and fill water buckets from the big portable tank. When Daniel was still in diapers, he carried chickens from the killing cones to the scalding tank. I cannot begin to enumerate all the normal farm hazards we need not worry about, and on the other side, all the positive things we can let the children do.

The earlier they can start, the deeper their commitment to the family farm enterprise and the sooner they develop the common sense necessary to carry them into the big jobs they will do as they grow older. If our farms do not provide opportunities for children, they will occupy their time sitting in front of the TV, playing little league, going to the mall and becoming video-game junkies. Then when they turn 13 and can really be helpful, we've lost them to all these other interests and the farm is not on their list of fun things to do. This is a great tragedy, and one of the most significant losses on the average American farm.

In addition, our farm must allow for emotional enjoyment. The conventional day-after-day grind, the stereotypical drudgery of farm life, is devastating to a family. We must have slack times if we have busy times. We must have breaks from the routine. And we cannot work under the lights very often.

When folks think that our models require too much time, I simply smile and respond: "Well, I don't have lights on my tractor." During blizzards in the winter when everyone else sees their chore time triple and the animals suffer due to stress, we just walk out to the hay shed, throw down some hay, take care of the layers in the greenhouse, and spend the day reading or writing a book. Reading to the children occupies a lot of our winter time. Since we do not have a television, we have time to do other things.

One other note about being family-friendly: seldom do we have a sick animal. Sick animals devastate children. Certainly anyone who grows anything will have something sick from time to time but it should be a rare thing. Generally, the plants and animals are healthy and functioning the way they are supposed to. Administering chemicals and pharmaceuticals all the time gets old real quick, and nobody likes to do it.

Our family-friendly farming model provides a safe place for children to be, meaningful work at an early age to develop discipline, responsibility and team loyalty, and an emotionally rewarding production model.

Home cooking instead of processed food. About 50 percent of all meals in America are prepared outside the home. This includes frozen pizza, TV dinners, restaurants and delis. The trend is going up, unfortunately.

Isn't it amazing that at the same time surveys show Americans are increasingly concerned about the safety of the food, we are entrusting more and more of its production and preparation to entities we don't know? Talk about intellectual schizophrenia.

And processed food is expensive. Farmers everywhere talk about a box of corn flakes returning less to the farmer than to the cardboard manufacturer. "The package is worth more than the food," the farmer complains.

Roughly 30 years ago, the farmer received 35 cents of every retail food dollar. Today his share is less than 9 cents and will soon hit 8 cents if present trends continue. Much of this, of course, is the centralization and vertical integration of American agriculture. But much reflects changes in buying patterns. People buy fewer raw foods like ground beef and potatoes, and buy more processed foods like noodles, bread, and entire meals.

The bottom line is, of course, reduced quality of the food because it is farther from the soil, not as fresh or local, and increased cost to the consumer. While certainly a market for prepared foods exists, we think it is better to encourage consumers to see the nutritional and economic value of home preparation.

Buying premium Idaho baking potatoes for 9 cents a pound sure beats paying 92 cents a pound for the same thing already sliced into french fries and stored frozen in microwavable bags. I saw precooked bacon in the supermarket the other day in 4 oz. boxes at $20 per pound! What's amazing is that people are actually buying that stuff.

Clean, nutritious, personally-inspected food. People's personal involvement with their food needs to be encouraged. I can't imagine why folks who see how many things get botched up by the government would entrust the safety of their food to that same institution.

Honestly, the government can louse up just about anything it sets out to do, in such creative ways you or I can scarcely imagine it. They set out to care for widows and orphans with something called Social Security, and end up with a boondoggle that will well-nigh bankrupt the country and that has grown into a retirement entitlement for all. They set out to reduce manure going into waterways and build manure lagoons that cause more contamination than if they had never been built. They set out to protect the American eagle and end up absconding a farmer's tractor and his livelihood because the fireline he ripped in as a last-ditch effort to save his home upset a nest of endangered mice. They set out to reduce filth in huge slaughterhouses and put all the neighborhood mom-and-pop shops out of business. The list is endless, and it doesn't matter whether you're a liberal or conservative, you know about bureaucratic corruption and stupidity.

Why would any of us entrust the government with protecting our food? Do you think all the huge outfits that receive government oversight are dedicated to turning out the cleanest, most nutritious food? Of course not.

The sooner we involve people with their food and show them there is something better than Twinkies and Pop Tarts, cardboard tomatoes and cellulose apples, greasy spareribs and pale eggs, they will realize that the "system" is totally rotten. We cannot get an educated, proactive populace as long as we have an agriculture so far removed from end users that they think milk comes from a jug and fish sticks swim around in the ocean.

Nothing is as clean and nutritious as a customer-inspected facility and business. When the relationship marketer must pass the scrutiny of the discriminating patron, who personally looks over the entire operation, you'd better believe not one in a thousand would risk doing something unscrupulous. But a huge entity with a lot of "No Admittance" doors, with a bank of Philadelphia lawyers and millions in product liability insurance to protect it can hoodwink, smooth talk, and weasel out of anything.

Let's restore integrity in the food system by getting involved with it, personally.

Non-embarrassing farm incomes. Why should a farmer be paid less than an average white collar city worker? Is he any less important? While I do not advocate for a minute some sort of arbitrary government parity program, or heavy-handed guarantee to make sure the farmer earns a living, neither do I think it is inappropriate for a farmer to price his wares in a way to return $20-$30 per hour for his time.

You are only as valuable as you perceive yourself to be. Often, this issue comes down to a self-worth, or self-concept view. How valuable do you think you are? How farmers can work themselves to death for pauper's wages is beyond me.

Dad always used to say: "You might as well do nothing for nothing as something for nothing." In other words, if you aren't getting any return, why do it? As an accountant, he saw farmer after farmer slaving away from dawn to dusk, never taking a vacation, and barely staying financially even.

I've often asked factory poultry house owners how the birds are doing financially. The common response: "Well, they're paying for the house mortgage." I'm sorry, but why go to all the effort of building a factory confinement facility if all it will do is pay for itself by the time it falls down and needs to be replaced? I can think of plenty more enjoyable things to do than pick up dead chickens and walk through ammonia-stenched houses. I'd rather hit myself on the head with a hammer.

Honestly, looking at the way many of us farmers value our worth, you'd think we were masochists. We just love to enslave ourselves to dead-end vocations just for the thrill of being called a farmer. I'm here to tell you life is way too short for that.

And the attitude this type of mentality and "unquality" of life produces is not one that is interested in producing the best milk, the best carrots, the best pork or the best eggs. The farmer is reduced to a robot, going about her chores without vision, without enthusiasm and certainly without creativity.

In order for us to have the vision and enthusiasm to produce better food every year and to think long-term, we must have a return that is highly rewarding. Money is not everything, but it is something. It is certainly a motivator and we do not apologize for demanding from our patrons, through the prices of our farm produce, enough money to hold our heads high on any downtown street. Any consumer who would deny that level of remuneration to folks who produce her food certainly doesn't have much respect for what goes into her stomach.

Emotionally exhilarating lifestyles. Go to any conventional farmers' meeting and you'll see a bunch of older men sitting around grouching about how awful everything is. They complain about the stranglehold of the huge corporations, from processors to fertilizer dealers. But of course every Monday morning they rush pell mell to patronize these entities they view as their enemies. They're all wearing caps emblazoned with the logos of these nasty corporations.

They think the environmentalists are out to destroy them, but then rush out and do things like build confinement factory houses and stink up the neighborhood, fueling the fire of their perceived enemies.

But our kind of farming provides emotional exhilaration. It's wonderful for the children because they can be part and parcel of the operation. It provides a financially rewarding income.

The seasonality allows for downtime, a change of pace. We can go on vacation, spend hours in the winter just sitting by the fire reading books and playing games. The excitement of discovering better ways to grow animals and plants so that we produce high quality food is truly exciting.

Wouldn't it be awful to get up and have to do the same thing every day? Or how about coming to January 1st and knowing you were going to do basically the same thing this year that you did last year? Armed with a list of improvements, new projects, additional plants or animals we want to try, and always knowing we'll develop relationships with brand new, wonderful folks, we anticipate the future with proverbial "thrills, chills and excitement."

The bottom line of all this discussion is that a farming enterprise succeeds or fails primarily on its philosophical underpinnings, not on hard-core how-to information. Whether you use Tipper Tie aluminum electric fencing or high tensile steel galvanized is no big deal. Whether you grow Better Boy or Rutgers tomatoes, the important thing is to have a farm model that jives with the right philosophy.

I have seen too many wanna-be farmers get mired down in unrewarding models and unrewarding policy because they deviated from the right thinking. In fact, many would-be farmers never sit down and articulate their philosophy like I have done here. If you do not clearly understand what you believe, you will be subject to all the peer and agribusiness pressures to deviate from the path I've outlined here.

Believe me, deviating from these philosophical moorings will guarantee shipwreck — maybe not immediately, but slowly and surely the farming venture will fail. If you opt for decentralization but, in your quest for new customers, sell too cheap to make a profit, you will fail. If you sell at a premium but do not produce a higher quality item, folks will discover the scam real fast and you'll be out of business. If you make piles of money but never take a break, the family will get tired of it, your spouse will get cranky, and it will all crumble. Each of these segments has a part in the success or failure of the entire operation. Think about them. Inscribe them in the recesses of your mind so they can be used any time to discern what is right or wrong.

Before you acquire your first plant or your first animal, before you put a dime into your farming venture, make sure you have your philosophy right, and you will be amazed at how much simpler your decision-making will be. While it will not guarantee success, it will help keep you on track, and will go a long way toward making sure your thinking and working are progressing toward success.

Do It Now

I wish I had a nickel for every time I've finished speaking at a conference and someone asks: "How do I get started?" My response: "What are you doing NOW?" It makes no difference whether you are 14 or 64. Your response to that question will tell you a lot about how you will do as a farmer.

First of all, I can't offer a five-minute answer to the question. At least now I can offer a book on the subject.

Secondly, everyone's circumstances are unique, and will necessarily require a different working-out of the farming enterprise. Variables regarding type of plants or animals to raise, location, experience, abilities, and on and on will govern how you actually proceed from where you are today to where you want to be.

Fundamentally, you must look around and ask: "What am I doing NOW? What can I do NOW?"

You see, most folks I've dealt with, who really want to farm, have the notion that if they just had some land, or if they just had more land, they could farm. It's as if an elusive something — land, equipment, buildings, markets — is always just beyond their grasp and they are just stuck until they can acquire that magic "thing."

Farming is not a "thing." It is a life, and a business. Plenty of farmers are awaiting the magical "thing" to become profitable. We make all sorts of excuses to explain why the farm isn't doing very well and why we need to drive into town every day for that steady paycheck.

Either said or unsaid, the mentality is:

- "If I just had that additional acreage over there."
- "If I just had that guy's tractor — you know, the one that runs — instead of this one."
- "If my kids were more interested."
- "If my wife was more interested."
- "If I just had that guy's bull instead of this old scruffball."
- "If I just had a better education."
- "If my equipment were newer."
- "If my corn were taller."

This thinking permeates the farming community, the people who already have a couple hundred acres but can't make a go of it. Nothing about owning land or being born into a farming family inherently removes the creative capacity to find excuses for our not being where we'd like to be.

To the adolescent, perhaps an FFA member, I would ask about current activities. Are you spending weekends watching TV and going to the mall, or working on a nearby farm — at NO pay — to learn about it? The truth of the matter is that many folks who say they'd like to farm really aren't willing to devote themselves to it.

I've known just a handful of youngsters who really had their heads screwed on straight, and who devoted their attention to spending time on farms, to talking with farmers, and who consequently developed contacts that carried them a long way in their agricultural pursuits. You've just got to decide whether you want this life badly enough to forego weekend indulgences, summertime swims and recreation, to begin something in agriculture NOW.

Chances are if you have no desire to grow anything now, you probably never will. You can grow something, even if it is a plant in a windowbox. But most folks can get access to a few square feet of ground, even if it is a spare flowerbed around an apartment complex.

Quit mowing the lawn and turn it into a backyard market garden.

If you're more concerned about what the neighbors will say when you convert your lawn into garden beds than you are about getting started in farming, you're too peer-dependent and would not do things differently enough to succeed even if you did have a larger acreage. Better to find out now than later.

Almost anyone with a tool shed could raise a few rabbits to harvest for meat. You can eat some and sell some. If you do away with a rooster, you can quietly run a pen of chickens around the yard, supplying neighbors with excellent eggs. You can hone your marketing pitch, develop your farming identity, and gain valuable production experience right in your backyard.

Don't worry about zoning ordinances — they're just there for when people complain. If you give gifts to the neighbors and turn them into patrons, keep things odor-free and tidy — and home — chances are you can develop quite a little menagerie before anyone notices.

I know a fellow in New York City who grew several batches of broilers in his garage — and dressed them there as well. He developed quite a following and is now patronizing a pastured poultry grower about an hour away and delivering them to his customers. That way, if and when he finally does get access to some land, he will have his clientele already in place.

If you have no land, you can purchase grow lights, get rid of the TV, and turn the family room into a multi-tiered garden, even if it only produces fresh salad greens for your table.

More folks than not know someone who can get access to some land. Your church no doubt has an elderly retired farm couple with land they are just renting out to some other farmer to keep the taxes low. Generally the rent is atrociously low, and you could go in with a market garden, some pastured poultry, or a rabbitry and get a huge start.

I would encourage you *not* to buy land. Land ownership is *not* the place to start; often it is the worst place to start. We are responsible for what we have, not for what we don't have. The world is full of folks who inherited farms and then went belly-up. Land ownership has absolutely nothing to do with successful farming, any more than owning a business makes a man a successful businessman.

Certainly I am not opposed to land ownership, but at the outset I think it necessary to debunk the myth that it is the prerequisite for farming. The back-to-the-land hippies soon learned that there was more to successful homesteading than building a dome home, planting two rows of adzuki beans, free ranging some laying hens and lying on a yoga board 30 minutes a day. Until you know how to make money on one acre of land, you won't know how to make money on two acres. Owning land is not the most important prerequisite to becoming a farmer. The how-to is what is most important.

The most salient requirement for farming, as I see it, is experience. Acquiring information comes in a close second, but nothing beats experience. Unfortunately, if our paradigm for acquiring farming experience requires land ownership, then chances are that neither will occur.

Actually, for folks who want an agriculturally based business, the best investment may be an inspected kitchen. In most cases, for under $10,000, a separate kitchen can be constructed that will meet inspection requirements and allow you to sell baked goods, many canned goods, and combinations like pot pies. Perhaps you and some buddies could resurrect a community kitchen or cannery. Such a federal-inspected facility would open up all sorts of opportunities for value added marketing.

That amount of money could easily put you in a self-employed cottage business but would scarcely buy you three acres of land, and certainly not a house, were you to relocate to a "farm." Once you have a thriving customer base for your kitchen items, then you can think about vertical integration to increase your gross margin.

For example, if you are making pound cakes, you could put in a flock of pastured laying chickens to produce the eggs preferred by gourmets the world over for flavor, texture and color. Each pound cake requires half a dozen eggs, remember. That way instead of getting a dollar a dozen you can get equivalent to two dollars a dozen because you are marketing them through pound cakes: this is value adding.

From a "getting started" perspective, acquiring land before experience or customers is often getting the cart before the proverbial horse.

Perhaps you can relocate without spending money. If you have no square footage where you live, consider all the want ads for "elderly person needs single or couple to live-in." With rising costs of elderly care and the fragmentation of the family, every community has several of these opportunities. What a wonderful ministry. And nearly all of these are older folks who have an expansive yard and garden area. Depending on the owners' biases, children can be an asset or liability for "getting in." Lively, well-behaved children can often be a real plus, but that is not the rule. Having children can often be an asset when answering other "groundskeeper" want ads for absentee owners where housing is provided in exchange for security and maintenance. In these cases, owners view children as a stabilizing influence in the home. These opportunities allow you to get onto at least a yard without buying land or housing.

One way to learn what people really want in your area is to do some marketing for alternative farmers nearby. You could develop the customer base, hone your marketing skills, and find out what is currently being supplied and what folks are wanting that is not being supplied. You may even modify what you thought you would grow based on this information.

Participating in the local "tailgate" farmers' market can give you invaluable experience without a big investment. Most have low-cost spots available for small producers. You can learn about production, marketing, what items are in short supply and what are in oversupply. Meeting the other vendors can offer great networking opportunities to find land or other farming opportunities. Just visiting with older farmers, gleaning advice from them, while you're waiting for your next customer, can yield dividends as you pursue your farming venture.

Consider apprenticeships, of course, both remunerated and gratis. If you realize that you are getting your farming education, you can do anything for awhile.

Instead of working out at the local health club or joining the sports league, spend your free time putting in some garden beds, putting in fence posts on a friend's farm or peddling your produce. Whenever I hear someone say they just can't get going because they don't have "time or money," I'm tempted to finish the statement with "for that."

We generally have time and money for what we really want. What are you willing to sacrifice in order to have your dream? Do you have to have that horse or that hobby car that takes time and parts? Until this idea of yours becomes your passion, you will not be focused enough to do what it takes to realize your dream.

The list of things you can do NOW is endless. At the least you can attend your state's sustainable agriculture conference and begin entering the loop you want to join. Instead of making excuses for why you can't farm, begin doing something NOW. Perhaps you will find you really don't want to. That's fine. Better to find out now rather than after you've sunk your life savings into it.

Again, lest you forget, remember that most bankrupt farmers inherited their land. Having land has nothing to do with farming success. It may be years before you actually have your land, and if you wait until then to really start in earnest, think how much catching up you'll have to do.

It's not as important WHAT you do, as long as you do SOMETHING, and as long as you do it NOW. Do something, even if it's wrong. You'll always learn more from your mistakes than you will from your successes. Go ahead, take the plunge, and get started NOW.

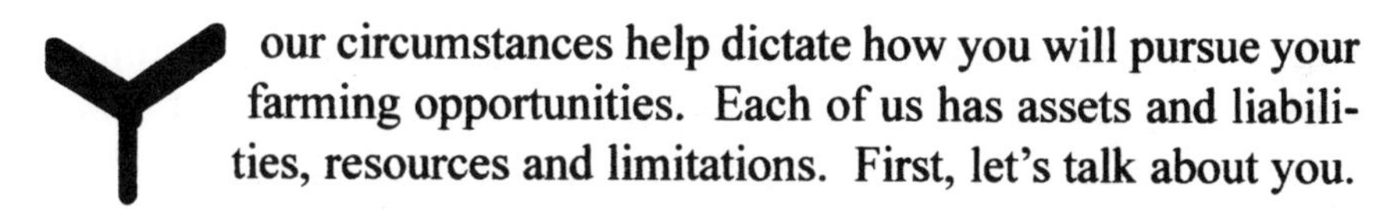

Your circumstances help dictate how you will pursue your farming opportunities. Each of us has assets and liabilities, resources and limitations. First, let's talk about you.

Believe it or not, age is perhaps one of the most important factors in how you pursue a farming enterprise. As age increases, the following things happen:

Physical stamina lessens. Most 60-year-olds cannot put in the same day's work that a 25-year-old can. You tire quicker.

Emotional stamina lessens. As a young person, you can roll with the punches a little easier. After a setback, or a really long day, you can go to sleep knowing that tomorrow you will be rested and ready to go. An old person goes to bed knowing he will wake up in the morning stiff and sore. Knowing that it's not going to get better is a downer.

Wisdom and experience increase. Older folks often contribute more by teaching and counseling than they do by actually doing the physical grunt work. Young folks need the wisdom of elders as much as elders need the vision and can-do attitude of young people.

Responsibilities increase. A young single is entirely flexible to experiment and be creative. Marriage adds another person to filter your ideas, although often this stimulates creative problem solving. Adding children changes the streamlined flexibility, and the freedom to risk lifestyle and income. This cycles, however, so that after children are gone from the nest, flexibility increases again.

Sociologists generally agree that men and women go through the life cycles differently. This has nothing to do with inferiority; it has everything to do with what you can do right now at your age. Men tend to begin running out of steam in their late 40s and early 50s. Note when the average corporate executive gets turned out to pasture — if he survives to 55, he's unusual.

Men pour their hearts and souls into those young years and, as they age, tend to have fewer new ideas and less drive to implement them. Certainly all of us can think of exceptions — Thomas Edison, for example — but they are rare. The sheer thinking flexibility required in a learning curve, a new enterprise, requires a man to be in his earlier years. He simply cannot adapt as readily as a younger man. "You can't teach an old dog new tricks" was not coined as much for animals as it was for men.

A woman, on the other hand, hits those later years free of the children and almost bursting with creative, experience-synergized energy. She is ready to take on new projects, see the world, start afresh with this man who has shown himself to be successful. The man she married was unproven. Now he's proven he can conquer the world, and all of a sudden he hears: "Honey, let's do this and this and this." Before you know it, he's worn out and she wonders why. I'm not trying to bog this discussion down with marital problems, but I think it is important to understand where we are on the cycle of life as we embark on new ventures. There are times to sail off into uncharted waters and there are times to look around for a few other ships to go with us.

Age makes a huge difference when you read this book and say: "Hey, I'm going to do that. What a great life this would be, and

I have a real yearning to be my own boss, to be out in green fields and watch the bees and flowers, and have a quiet place to raise my children." A 25-year-old and 50-year-old will not come to the project with the same zest and staying power. Older folks should find a young partner with which to enter the venture.

After speaking at a conference, I usually get several men between 45 and 55 who come up to me and say: "My wife and I are trying to start a farm, but we just don't have the energy. How in the world did you get it all done when you started?"

My response is simple: "I was 10."

Dad used to say: "The only thing worse than starting late is starting never." But we must be content to understand where we are in the cycle of life and realize that any new venture with its inherent learning curve, its ups and downs, takes huge investments of energy and we haven't failed if we do not completely attain success. Perhaps your contribution to mankind will be to give the next generation a head start.

That is essentially where Dad and Mom were. They were already in their late 30s when they bought this farm, and they didn't have portable electric fence and useable chain saws and front end loaders. More importantly, they did not know a tenth of what we know now. How I wish they could have had a book like this when they started.

But they paid for the farm, did much conservation work, research and development, and paved the way for me to pick up the baton and run with it. They were not failures for being unable to fully enjoy the financial success of the farm. The best they could do was to enjoy the vicarious thrill of watching the next generation fly. Any parent who watches a child succeed knows that thrill. Whether she's winning the spelling bee, hitting a homerun or bringing home an achievement award in 4-H, her parent's exhilaration is every bit as real as the youngster's.

And this brings me to the next big factor as you assess your resources: finances. Fortunately, to implement many of the ideas in this book does not require large outlays of capital. Most of the ideas described here require creativity, savvy and raw human energy. That is not too difficult for a youngster who hasn't been tempered with age. But I am well aware that many, many of the people interested in starting a farm are in this middle-age bracket, and it is to you that this discussion is critical. The advantage of coming to this desire, or recognizing the passion to farm, as an older person is that often you have more financial resources. You have equity in a house, perhaps some retirement funds, some investment capital and a savings account.

The money you put into the farm will yield far greater dividends if you partner with a young person than if you try to do it yourself. For you over-45-year-olds, this may be the most important statement in this book. Young people, under your wisdom, can turn more of your money into profit than you can. Your big earning days are coming to an end, and it is time to appreciate that you will do far better sharing your investment with a youngster than you will trying to do it all yourself.

I know a dear couple, good friends, who fall into this age bracket. After their second year of full-time farming — doing most of the right things — they were extremely depressed. The workload was killing them. At their age, and with their children gone, they did not need to come in bone tired every night. They were doing about 2,000 pastured broilers at a net profit of about $5,000. The birds were giving them a wonderful return, but the workload was just too stiff. They found two young men and I counseled them to split the profits 60-40 with these fellows. That way, the older couple, who have the advantage of community respect and reputation, can devote themselves exclusively to marketing. These two youngsters could easily run them up to 8,000 birds, which would be a profit of $20,000 (8,000 X $2.50 per bird). At a 60-40 split, the young folks would get $12,000 and the older couple $8,000. Their income would increase by $3,000 and they wouldn't have to do any of the heavy fieldwork. This would be kind of like a pension.

Our apprenticeship program is working along these lines.

Apprentices who want to come here and learn receive better than a college education in 12 months and earn income while doing it. Our long-range plan is to break off our enterprises as self-maintaining profit centers for the next generation while we receive royalties. The next generation gets the benefit of our experience and capital while we get the benefit of their energy and curiosity.

As we age, we want to see our physical workload continue to decline. But in order to get "good help," we must be willing to share more than a token amount with the labor pool. Farmers who treat their children like slaves will not see the next generation desire to work alongside them. We have always paid our children big money for their work and this is one of the best insurance policies for sustained labor.

Often middle-aged folks desiring to "make the break" to farming have been successful elsewhere and have money to put into the project. I cannot stress enough that you would do far better putting that money into a conscientious young person, either one of your children or an acquaintance. They can take your investment, if their pay is commission-based, and generate far more income for you than you ever dreamed possible.

If middle-agers would put the same time and energy into finding young partners that they do in finding the right tractor, the right barn builder, the right seedstock producer, they would find their endeavor more rewarding and more lucrative. The primary reason is because both parties bring to the relationship their best assets and do not need to use their liabilities. The older folks bring their experience, their community contacts and their financial security but do not have to hammer every nail and saw every board; the younger folks bring their physical prowess and their youthful enthusiasm without risking major mistakes having to borrow money.

It's a win-win partnership, operating in symbiosis so that the whole is worth more than the sum of the parts. The older person must be content with where he is in life and extend a magnanimous helping hand to the youngster; the youngster generally will respond in kind if he thinks he's getting a fair shake.

Most of these relationships, whether familial or not, break down because the older folks do not extend enough freedom and remuneration to the youngsters. When the youngster has to ask for money to buy a Coke, or ask if he can use a board off the lumber pile, he won't last long. He may last a couple of years, but he will tire soon of his bondage and break out. Older folks must release, must allow the next generation to spread wings and fly.

The lion's share of capital in agriculture is held by old folks (the median age of the American farmer is now 58) and the young folks, seeing no future on the farm, have left. This is tragic, and it need not be. One of the greatest blessings of my life was having a Dad who not only let me try things, but who made a conscientious effort to hand the reins over to me.

From the day he knew I wanted the farm, he eased the control over to me. He began weaning the farm away from his financial backing; he made fewer and fewer decisions. That made the transition, when he finally passed away, simple and efficient. The farm showed not a sign of a bobble — it just kept on steadily climbing.

Somehow, older folks must dig down within their souls and release control, release remuneration, to the next generation. Men, if you do not know whether or not you are doing this, ask your wives. They know.

If you are a youngster reading this book, I have tremendous faith that you will succeed. You may or may not want an older partner. It would be helpful, but I know how hard it is to find older folks who can be persuaded that a farming enterprise can be successful. Too many of them grew up pulling weeds and hoeing corn and gathering eggs at pauper's wages and couldn't wait to leave the farm. Most farm children are still not making anything and can't wait to leave the farm. But it need not be.

Perhaps nothing illustrates the timelessness of this problem better than a quote from C.C. Bowsfield's little 1913 book, *Making The Farm Pay:*

"The average landowner, or the old-fashioned farmer, as he is sometimes referred to, has a great deal of practical knowledge, and yet is deficient in some of the most salient requirements. He may know how to produce a good crop and not know how to sell it to the best advantage. No citizen surpasses him in the skill and industry with which he performs his labor, but in many cases his time is frittered away with the least profitable of products, while he overlooks opportunities to meet a constant market demand for articles which return large profits.

"Worse than this, he follows a method, which turns agricultural work into drudgery, and his sons and daughters forsake the farm home as soon as they are old enough to assert a little independence. At this point the greatest failures are to be recorded. A situation has developed as a result of these existing conditions in the country which is a serious menace to American society. The farmers are deprived of the earnest, intelligent help which naturally belongs to them, rural society loses one of its best elements, the cities are overcrowded, and all parties at interest are losers. The nation itself is injured.

"Farm life need not be more irksome than clerking or running a typewriter. It ought to be made much more attractive and it can also be vastly more profitable than it is. Better homes and more social enjoyment, with greater contentment and happiness, will come to dwellers in the country when they grasp the eternal truth that they have the noblest vocation on earth and one that may be made to yield an income fully as large as that of the average city business man."

This is one of my favorite passages in all the agriculture books I've ever read. It could easily have been written yesterday. Truth is timeless, and this treatise, nearly a century old, illustrates that the more things change, the more they stay the same.

In addition to looking at finances, you need to survey your knowledge and experience. Obviously the more you have of either or both, the more aggressive you may want to be at jumping into the farming enterprise. The less of each, the more conservative you will want to be.

Knowledge and experience are best gotten at the hands of successful practitioners. Probably the best way for a young person to start into farming is to work with an older farmer. If you have a relative or an acquaintance who farms, spend all the time there you can. It takes time to develop common sense.

What many of us country boys call "horse sense" can't be learned in a book — not even this one. The dilemma that I see in the alternative agriculture movement is that most of the young people who grow up on farms and therefore have the common sense to operate machinery and handle animals are not open to new ideas; the city youngsters that are open to new ideas do not have the common sense to operate a farm.

Although I differ dramatically from mainline conventional farmers, we still do many of the same things. And even if you spend time on a conventional farm, you will learn much about handling machinery, handling the soil, animal husbandry, cowpies and mudholes. Just tune the old codger out when he complains in his beard about there being "no money in farmin'." Take the worthwhile and leave the bad.

Next, look at your location. For that matter, look at your current farm base. In his fabulous book *Backyard Market Gardening,* Andy Lee points out that a back yard is a wonderful first farm. Do not dismiss anything as being too small. One of the biggest deterrents to would-be farmers is that they have this idea that a farm is supposed to be so many acres out in the country, with woods and ponds and cows all around.

Nothing could be further from the truth. Your biggest asset may be that you live smack dab in the middle of a city. You're right in the middle of consumers, ready to buy something local and fresh.

I know a fellow who started his farm from a condominium. He knew a parcel of vacant land between his condo and his office, so he just interrupted his commute between home and work to take care of his chickens.

I know an elderly couple in Wisconsin who live in the city and own a relatively large backyard. They built a portable chicken pen and in their first season raised 300 broilers in their backyard. Now they're planning to talk to the neighbors and see if they can fertilize their yards as well. The neighbor children love to see the little chicks and the model makes the prettiest lawns in the neighborhood. This is farming right in the city.

Finally, take careful assessment of the people in your life. Your spouse, your children, your parents, your in-laws, folks at work, friends at church — the list could be endless, but you get the idea.

Make a list of what each could contribute to the venture. Do they want to be mechanics or engineers who can help you design machines or tools? Do you know a good artist who can draw attractive illustrations on marketing brochures?

Perhaps you know a well-to-do person who, after reading this book, could be persuaded to partner with you. If you're older, do you know an aspiring youngster — a home-schooler maybe — who is honest and dependable who could provide some labor?

The people in your life undoubtedly will be your most important resource. Be sure to itemize the varied talents and assets each can bring to the venture so you will not overlook that resource. The problem is that once we're into the enterprise, we often overlook contributions others can make because, after all, we're being an "independent farmer."

Let me go on record right here that I am not an independent farmer. I am completely dependent on others — in a good way. The whole idea of community, of working together, presupposes that we help each other, and I am not afraid to ask someone for help. Most folks are tickled to death to lend a helping hand, but no one asks. Be quick to allow others to feel a part of your farming enterprise. Interdependence is strength.

People are starved for roots to grandpa's farm, for links to the land. You can provide that by sharing your needs, as well as your dreams, with them. Let them know they are welcome to help, and be sure to mention specific needs to the appropriate people. You would be absolutely amazed at how helpful people are if you but let them feel a part and register your need.

Of course, you will want to reward these helping hands with a dozen eggs, some garden-fresh tomatoes, and the like. This will ensure you a steady supply of helpful folks.

The bottom line is that as you assess your situation, see your circumstances not as obstacles but as stepping stones. Each individual's path to success will include a unique set of people, money and land. The mechanics will work themselves out as we devote ourselves to using the right stepping stones.

Good Enough

Many folks never get their farms off and running because they have to have everything figured out before they make the first move. You don't have to know all the ins and outs of a project before you start it. My heart aches for people who are hamstrung because they can't deal with imperfections or uncertainty.

My uncle was a top-level automobile manufacturing executive whose contribution to mankind was the disc brake. He described the early days of the disc brake project. The tool room fellows prided themselves in being able to produce calipers, one of the basic components of the disc brake, to within one-millionth of specification.

The competition was making them to about one-thousandth of specs. It was wonderful to have such perfect calipers, but the cost of getting them that close to tolerance made them too high-priced to be salable. Salesmen were frustrated and the tool room guys were pleased as punch.

The vice president finally called in the tool room guys and told them that he was proud of their manufacturing achievement, but if it made such an expensive product they couldn't sell it, it didn't make any difference how wonderful the calipers were, they'd all be out of business. He said: "Look, all we have to do is make them better than the competition. Good enough is perfect."

Our culture is cranking out technicians left and right. The

most common technician is a computer programmer. In order to function, a computer requires a certain procedural sequence every single time. If you mess up one little step, it won't do anything. The sequence must be perfect every single time. Any computer person who decides to start farming has a real uphill battle dealing with these thought processes. You can't just walk away from the office and think everything will be fine down on the farm. There are basic, fundamental thought processes you'll have to overcome.

While I use a computer, I am by no means proficient at it. People who really enjoy playing around with computers generally do not make good farmers. Not only can a computer consume your day just because it is interesting, it can also create enough marvelous spreadsheets and diagrams that, for small projects, you can easily spend several hours doing what you could do in a few minutes with a pencil and paper.

Computers work well when you're handling piles and piles of information. But the amount of information must justify going to the machine. For example, some people can't believe we don't have our 400-person customer base on a computer. I can hand-write addresses on all the envelopes in less than 3 hours; we only do one mailing per year. Do you know how long it would take to enter all those names and accompanying data in the computer? The second it's entered, it's outdated because someone moves, someone is added, or whatever.

Now, when someone changes addresses or we want to add someone or pull them from the customer box, we just throw away the index card, pull the index card or rewrite it. I can do that in a fraction of the time it would take to boot up the computer, get into the program and make these changes.

On the other hand, if we had 100,000 customers and a full-time data entry operator who always had the computer turned on to that program anyway, making those changes would be cheap and efficient via a computer database. We must be careful that we do not let our machines become the proverbial tail wagging the dog.

I can't believe how many programs are out there now to help

farmers make decisions. Ninety percent of them are just time sappers. Most of your decisions are not that difficult and require more subjective cogitating than a computer can do anyway. In the amount of time it will take you to come into the house, crank up the computer and get a meaningful diagram, you can just roughly figure the problem on a piece of paper or in your head and make a good guess.

A young fellow came to the farm for a few days to "see how you get it all done." When he got ready to leave, I asked what he discovered. He responded: "I found out how you get it all done: you just do it."

How profound. He captured the essence of our success. I run into people all the time who spend hours and hours figuring but can't seem to get anything done. Look, the chickens and cows don't care if the gates are crooked or the fence boards aren't perfectly square-ended. All that matters is function. We can get so bogged down on form that we never get around to function.

Who cares if the potatoes aren't in perfectly straight rows if you are just planting them for your family consumption? If they are going to be hand-hilled and weeded, it doesn't really matter. Now if you're in the commercial potato growing business and everything is being done by machine, straightness is critical. But the potatoes don't care if they're crooked; what matters is to get them into the ground.

So what if you miss some hay with the mower? So what if you overgrazed today and undergrazed tomorrow? The grass is forgiving; the cows can make compensatory gain. The earthworms can harvest the carbonaceous grass. The point is to get the hay mowed and the cows moved.

I'm not suggesting that the procedure is completely unimportant, but there must be a balance between doing things perfectly and doing things at all. I'd rather see someone doing things, even if they aren't perfect, than spending all day stewing and never getting anything done.

Here at Polyface we are strictly utilitarian, and this may be one reason why nearby farmers don't want to do the kinds of things we do. Ask the average person to describe the successful farm, and you'll hear about pretty fences, painted red barns, waxed green tractors and manicured lawns. Because people have a jaundiced sense of what 'success' looks like, they think the lean and mean, threadbare look of a truly lucrative farm indicates lack of care, negligence, and poverty.

We need our junk pile for all the exciting projects we're working on. We're building things and tearing apart things all the time. If you want to see beauty, I'll take you out to look at the animals, the grass, the ponds and the earthworms.

Photo 6-1. *A simple, crude chick brooder big enough for 300 at a time illustrates function over form. A few minutes and a few salvaged boards and we were in business.*

Too often people get so bogged down in appearance and having everything just right that they never get the basic project underway. Believe me, the pigs are much more interested in a feed trough than they are in whether or not that feed trough is perfectly square. The difference between doing a job functionally and doing it perfectly can be hours, and those hours add up to dollars, and those dollars add up to profitability.

Certainly nature does not require perfection. All nature requires is performance. Making mistakes is exciting because we learn from them. Some people are absolutely paranoid about making mistakes and having people come see a wreck. That's how we learn, and if that bothers someone, let it be their problem, not yours. Many people are paralyzed until every single question is answered. If you wait for that, you'll miss out on life.

Many would-be farmers from the computer mentality proudly develop and display spreadsheets about their farming options while the cows wallow in the creeks, fence posts rot off, and weeds take over the vegetables. You don't have to figure out every contingency, every possibility, before you make the first move.

I'm not suggesting planning is unimportant. I am suggesting that you can figure and figure and figure until the project never gets done.

Dad was a journeyman pattern maker in a large General Motors factory before he went off to fly airplanes for the Navy in World War II. These patterns were wooden models and molds for manufacturing metal auto parts. That kind of work required extreme precision. He was meticulous to a fault. He could easily have built dining room furniture, grandfather clocks and other close-tolerance wood creations.

I am exactly the opposite. I don't think I've ever built anything square in my life. We have a joke around here that I use special measuring tapes that only have marks on inches-nothing in between. One of my jokes on other people is when they come out to work on a project with me I'll grab their tape measure and exclaim: "Bill, what in the world are all these little marks on here? I never saw all these before. This must be one of those seconds that didn't get made right." Everyone has a good laugh over it. If we're out building something and I ask someone to get the carpenter's square, Daniel and the apprentices will hoot: "Whoa, boy. He's getting serious now. Watch out. This thing's actually going to be pretty."

Anyway, to his great credit, for all his ability Dad never complained about my 85-degree angles. If he had, I probably would have quit building anything and that one thing could have shut down the farm. But no, he encouraged and was glad the item worked. I know it must have caused him internal consternation, but he never showed it.

We use the "good enough is perfect" phrase around here all the time. It's a liberating, freeing concept. The important thing is to get started, to do something, and learn as you go. If it works, that's all that matters.

I don't care if you have 5 cows or 100 cows; you'll always have one you want visitors to see and one you hope stays out behind the bushes. No matter. The only way you know which animals will work is to have some that do not. As you learn those characteristics, you'll make better culling decisions. Don't feel like you have to go out and buy clone-looking registered stock to get into the cattle business.

Don't cripple your enthusiasm by stewing over whether to sucker or not sucker the tomatoes, whether to trellis or let them spread, or what kind of trellis if that's the best. Just enjoy the discovery process. Try some as many different ways as you can, and then watch the results. Until you get them in the ground in any shape, you'll not begin to know what is best.

A neighbor told me about an old-timer here who always had a picture perfect garden. The neighbor asked him if he planted his potatoes in the signs. Without missing a lick, the old-timer responded: "In the signs? I plant my potatoes in the ground."

Until the potato is in the ground, it doesn't really matter if it's in the right sign of the moon or not. I do think there is something to planting in the signs, but it's more important to get them in the ground.

Be encouraged; don't be paralyzed. Just go do it, and don't worry about what the neighbors will think. Don't worry about the local protocol. And don't demand perfection from your spouse or your children. Let them have some freedom to create and invent too. That's the way to stimulate folks around you to do things. Just remember, good enough is perfect.

For Beginning and Profiting . . . Lessons from the Lemonade Stand

One of the biggest needs of farms — indeed of all businesses — is a model that allows beginners to make a profit on a small scale.

We live in a time when most businesses do not realize a profit until several years, many 80-hour work weeks and stacks of debt payment vouchers after the initial opening of the business. Interest payments and market access are a daunting gauntlet for would-be entrepreneurs.

How many times have you heard someone say that you can't farm unless you inherit it, marry it or become independently wealthy? I used to believe that too, but not any more. If this statement were true, why are so many 40-year-olds who inherited their farms either going bankrupt or commuting to a town job?

Opportunities abound for folks to enter farming without the typical debt-financed or outside equity-financed start-up phase. Perhaps we all need to sit back and take a fresh look at successful child-operated lemonade stands.

DO SOMETHING YOU ENJOY. Kids love lemonade. Kids love to earn money. Kids usually enjoy people, and especially enjoy hearing compliments. Read any survey about why people do not enter farming, and you'll see words like "drudgery, dust, noise, smell, long hours, low pay."

Unfortunately, these descriptions are correct for 95 percent of American agriculture. But that simply means the models are flawed. Just because most people are doing it wrong doesn't mean you can't do it right. There is an entire alternative world out there that is as foreign to the average farmer as Raymond and Dorothy Moore are to the average school board member. Farming can be enjoyable.

If you don't like carrots, don't raise them to sell. If you don't like baking, don't attempt to sell bread. If you really like eggplants, maybe that's something you could try to market. Building a successful business requires tremendous energy, and if it's something you enjoy, that will help fuel your inner fire.

I don't raise anything I don't like to eat. At least if it doesn't sell, I can sure enjoy eating it.

INTIMATE EXPERIENCE. Kids are connoisseurs of lemonade. They know exactly what flavor, texture and temperature they like. Never try to market something you haven't tasted, felt, cooked and experienced personally. This is imperative if you are going to be knowledgeable and exuberant about your product — and if you're going to compare it to the competition. How many soybean growers have ever eaten tofu? I have run into a few soybean growers who make tofu, and their enthusiasm for the bean is infectious.

FIND A NICHE. What can't you get on a hot day in a residential section of town right out on the sidewalk? If you said, "An ice cold glass of homemade lemonade," go to the head of the class. You don't find lemonade stands next to Pepsi machines or blocking the Kool-Aid aisles at the supermarket.

What do people want that they cannot find in the mass market? The two things that are lacking at Wal-Mart are honest personal attention — ever try to get real information from an attendant? — and high quality. The buy-it-today-throw-it-away-tomorrow mentality permeates our industrial society.

And if you haven't noticed, we are living in the era of the niche. Designer this and designer that — our culture is finally beginning to rebel at industrial sameness and this is creating huge opportunities for creative thinkers. The reason a 6-year-old can operate a successful lemonade stand is precisely because it is *not* a supermarket.

Specialty products are everywhere. Uniqueness can be tied to geography, season, genetics or production model (i.e., "organic"). The worst thing you can do is what everyone else is doing, and yet that is what most farmers do. If it's not authorized by the extension agent, it's not doable. Look at your neighbors. Can you imagine a business that only tried to duplicate what the next business was doing? Do something different.

PRODUCTION/PROMOTION. If a youngster spends 6 hours manning a lemonade stand, probably only about 30 minutes were actually invested in production: the remaining 5½ hours were in marketing and promotion. Any successful farming venture will be similar.

Nothing has destroyed more farmers than believing the lie that if they will just produce enough, their financial future is secure. Amazingly thousands of farmers still believe it, as evidenced by the numbers signing up as contract growers for vertical integrators like poultry and swine conglomerates.

Production is only a tiny part of the equation. Any enterprise that generates a profit in the beginning will be weighted toward marketing and promotion because that is where the real money is.

RETAIL PRICES. The price per cup at a lemonade stand is just as high as that of a soda. The gross profit margin between cost of inputs and the sale price is extremely high.

When you're producing millions, the gross margin can be relatively low and you may show a profit. In a small enterprise, you cannot afford to sell wholesale because your overhead in equipment and labor make you uncompetitive. Furthermore, you simply cannot sell enough product to pay your salary.

That is why I never envision us selling low-grade lumber. In our area of western Virginia, average hardwood stands will bring the landowner $300 per acre on the stump. A medium grade log that brings $20 at the sawmill will bring $200 at the hardware store. A low grade board sells for only 30 cents a board foot, but good quality walnut, oak, cherry or ash will bring up to $2 per board foot.

If you are running a sawmill capable of cranking out 20,000 board feet per day, 30 cents adds up to quite a bit. But if your mill only does 500 board feet, you'd better stick with the $2 stuff.

The cost is in being there. A cattle producer marketing 100 calves has a huge labor advantage over the one producing only 10 calves. A tractor trailer moves material more cheaply than a pickup truck.

Small farms simply cannot generate enough dollars to live on in the wholesale market. That is why retail prices and direct marketing are critical to success.

PARTNERING RESOURCES. When the youngsters come in with their handful of money, they do not pay Mom and Dad rent on the kitchen, rent on the pitcher and user fees for the beater or refrigerator. They utilize underused resources and turn an almost pure profit.

An idle backyard and a million-acre ranch are equally underutilized and awaiting creative entrepreneurs. I know a retired couple in Wisconsin who live in the city. They built three portable chicken pens and raised 300 broilers in their backyard, netting several hundred dollars in just a few months. They did not need to charge off the cost of the land to their farming enterprise.

Perhaps you know someone with a copper kettle — how about making apple butter in your backyard? Our son Daniel tapped our three yard sugar maple trees and one on the neighbor's place, boiled down the syrup and made maple donuts for the farmers' market, turning the syrup into about $200 per gallon. Which brings us to the next point:

VALUE ADDING. As much as possible, take the product to a more refined level. That's all kids do in a lemonade stand; most of us have powdered stuff in the pantry at home. All these youngsters have done is converted it into something highly consumable.

Raw materials have very few buyers. How many people want to buy a cow compared to the number who buy hamburgers? How many people want to buy a chicken, compared to the number who buy McNuggets? How about a bushel of corn, compared to cornbread? Get the idea?

As a beginning farmer, you won't get very far taking a bushel of corn down to the grain elevator. But growing some specialty corn in your garden, milling it in the kitchen and turning it into fresh designer cornbread . . . Mmmm! I can hear the phone ringing now and the cash register not far behind.

DIVERSIFY THE PORTFOLIO. What do really enterprising kids do to enhance their lemonade stand sales? They add cookies or brownies, of course. The hardest part is getting the customer. Once the customer is in the door, add more items and he'll surely spend more money with you.

Rather than thinking about growing more carrots or beef or corn, add different items to garner more of your customers' dollars. Although no one item will be enough to make a living, the sum can be. This runs completely counter to the conventional agriculture wisdom of monocultures and single-item production. We even have poultry farms that grow only cockerels. Not only is that quite unnatural, but it guarantees the lowest prices for the farmer.

If you're making rolls, add jams and jellies. If you sell pork, add beef and chicken. We've found that we can get roughly four chicken customers to each beef customer. Our chicken sales have brought on more beef customers. The different items act symbiotically to make one plus one equal three.

ENJOY PEOPLE. How many times have we seen pictures of kids falling asleep at the lemonade stand? Be assured, when that happens, they have literally marketed themselves to exhaustion.

Enthusiastic selling takes energy. Energetic marketing is simply enjoying people. You're tuned into their body language, the hidden meanings between their actual words. You're trying to meet their needs, and that takes energy. Cherubic, exuberant young faces make people part with their money faster than anything.

Too many people want to farm to "get away" from people. That goal will doom any farm to failure. I literally do not know of a single really thriving farm, big or little, that is run by people who don't like people. I know many degenerating farms run by people who don't like people, who complain about those ignoramus city folks.

When you add people to the farming equation, the dynamics change. You will make completely different decisions because you will be thinking about what customers want instead of what the neighbors or the extension agent wants. You will be extremely careful about your craftsmanship and production/processing practices because you are seeing those customers face-to-face routinely, heightening accountability.

If you want to get away from people, don't try to succeed on a small farm. You'll need people to make it thrive.

BRAINSTORMING. Most lemonade stands do not come into being spontaneously. They are the result of children and parents sitting around talking about ways to earn money or neat things to do. They are born out of familial communication and brainstorming sessions.

Don't be afraid of your dreams or ideas. Vocalize all the potentials; write them down. Whether it's sources for financing, ways to approach the landlord about using that condominium backyard for something other than a lawn (including baking her a pound cake and taking a bouquet of window-box-grown flowers), querying your acquaintances about what they'd like to buy that they can't get in a supermarket, focus attention on where you are right now and on what you can do right now.

Very seldom is success spontaneous. It comes after much hard work and growing pains. If you think you need to own land to farm, forget it. I know folks who live in condominiums who have thriving pastured poultry enterprises on land down the road. I know a fellow in New York City who grew chickens in his garage and sold them to folks at work. Plenty of successful vegetable enterprises started by someone parking the lawnmower and converting the yard into a garden.

I can't begin to list all the possibilities here, but I can assure you that whatever your circumstances, whatever your situation, someone has overcome similar obstacles in their quest for dream fulfillment.

Examining Your Ideas

Recipes for Failure

Here are what I consider the ten most common pitfalls among farmers. Of course, these are principial and have applications to many things besides farming. But my applications will be specifically for farming. These are listed in order from those that are least devastating to those that are most devastating.

#10 BEING TOO INDEPENDENT/SELF-SUFFICIENT. Are you the kind of person who doesn't like to ask directions when you get lost on a trip? Here you are, not quite sure where you are or where you're going. Somehow you can't quite make sense of the directions you were given, or you just can't figure how those two roads connect in the crease of the map. Anyway, you're turned around and feeling a little confused.

Do you try to find the right way by just driving down different roads, discovering different options, hoping you'll stumble on the right one? If you do, you're probably not ready to farm. For one thing, this blind stumbling around mentality is not only extremely inefficient but it's emotionally draining. To succeed, you'll need both efficiency of action and emotional exuberance.

I confess that my family knows me as one who, at the slightest hint of bewilderment, stops and asks. If I get turned around on a college campus only a quarter mile from where I'm supposed to be,

I'll roll down the window and ask a student where such and such a building is. I hate to be lost.

Teresa and I laugh about the time we went to a public meeting put on by our congressman. It was held in an elementary school in another town. We knew the general area but had never seen the school. We got near where we thought it was about dark and I stopped at a gas station to ask where this school was.

I'll never forget how the attendant looked up, incredulous, and just pointed next-door. "There," he said. He probably laughs about it to this day. But you know what? If I had driven past it, I would have wasted time, gas and emotional energy.

The romantic notion that you are going to go out into the wilderness and carve out a livelihood that your spouse and children will savor, all by yourself, is simply nonsense. That is why in what we romantically call the pioneer days people had barn raisings and house raisings. Nobody was naïve enough to think he could do it all on his own.

The amount of skill, craftsmanship and information you have to know simply can't be self-taught without economic ruin. Oh, you may be able to subsist, but this book is not about subsistence farming. This is about making a decent living with enough money to give away to charitable causes.

Be quick to ask advice from neighbors and put attention on helping them in order to get them to help you. Seek counsel. Spend time accumulating how-to skill and wisdom. You'll need every bit of it as you get started.

#9 IMPATIENCE. In a nutshell, this is wanting too much too fast and then getting frustrated too soon. Anything worth striving for is worth waiting for. I've watched folks routinely give up right before the breakthrough.

Remember the normal business curve first slopes downward into the trough of despair before climbing up to the pinnacle of success. Studies show that most millionaires do not earn extremely high incomes. They only earn maybe $70,000 per year but they average 57 years old. They save aggressively starting young and staying the course until finally their net worth exceeds $1 million.

Generally anything that comes too easily and too quickly will not stay with you. Something will happen that makes it go sour. This is especially true in marketing. I can't tell you how many times a market has opened up for us that we thought would put us on Easy Street but never materialized. Instant euphoria will generally turn into later frustration.

If what you're doing doesn't work, keep tweaking, fine tuning and adjusting until it does. The way you plant the tomatoes, the way you build a chicken pen: these plans are not carved in stone. Be willing to take the bumps along the road as you gradually move toward success. I don't know how many people I've watched get production up to nearly 1,000 pastured broilers, right where I know it will really turn a big profit, and then throw in the towel.

One of my favorite stories is about Thomas Edison, as reported by his son Charles:

> "It is sometimes asked, 'Didn't he ever fail?' The answer is yes. Thomas Edison knew failure frequently. His first patent, when he was all but penniless, was for an electric vote-recorder, but maneuver-minded legislators refused to buy it. Once, he had his entire fortune tied up in machinery for a magnetic separation process for low-grade iron ore — only to have it made obsolete and uneconomical by the opening of the rich Mesabi Range. But he never hesitated out of fear of failure. 'Shucks,' he told a discouraged co-worker during one trying series of experiments, 'we haven't failed. We now know a thousand things that won't work, so we're that much closer to finding what will.'"

#8 NOT ENOUGH 'WHY' AND TOO MUCH 'HOW.' This deals with what I call the "untouchables" in a person's thinking. Or, as ranch consultant Stan Parsons likes to say: "We've become extremely good at hitting the bull's eye on the wrong target."

We've become adept at growing 150-bushel corn without asking the question why we should be growing corn. We've become proficient at matching exact drugs to precisely diagnosed bacterial infections in concentration camp poultry houses, but we fail to ask why we are raising chickens in concentration camps.

I suppose if the truth were told we all suffer from this "sacred cow" syndrome to an extent, but you need to minimize it. For example, I know a dear wanna-be family who purchased their dream farmette and most of the ten acres was already highly developed with espaliered dwarf apple trees. It turns out that these hyper-bred trees are fragile, disease-prone and fertilizer-dependent. The literature says that when everything is just right, they are downright lucrative. But this family didn't need something that only performed one year out of five. I was amazed at how much they knew about what sprays to use, and how many hours it took to mow and weed one row. These folks were experts at the "how," but in reality the best thing they could do might have been to bulldoze the stupid trees and do something else.

This same dilemma is common with folks who buy a picturesque place with a bank barn, those lovely 150-year-old structures with second-floor entrance and horse stables underneath. The idea was to enter on the second floor with hay and bedding material, then pitch everything down into the first floor where all the animals were housed. Of course, these were built when farm labor was plentiful and before much farm machinery. Cleaning out all those little bedded stalls by hand is a big job. Today, this building is as dysfunctional as they come and the thought of tearing it down is like asking someone to run through a fire.

Perhaps you're familiar with the folk tale about a lady who always cut off the end of the ham before putting it in the oven to cook. Finally her daughter got big enough to notice this every Christmas and asked her mother: "Mommy, why do you always cut the end off the ham before you cook it?"

Mother replied: "Because my mother always did it. I don't know why."

When Grandma came for Christmas dinner, the little girl asked her: "Grandma, why did you always cut the end off the ham before you cooked it?"

To which Grandma replied: "I don't know, dear, except that my mother always did."

Well, that mother wasn't around anymore to ask but there was one brother, an elderly bachelor uncle in his 90s who could remember, so they went to ask him.

"Uncle Mac, do you remember why great-grandma always cut the end off the ham before she cooked it?" asked the little girl.

Uncle Mac stared dreamily into the hills, as if trying to remember back that far, and then chuckled as the reality of the question and answer struck him. "Well, honey, " he said, "we'uns didn't have a pan big enough for the whole ham, so Mama cut a piece off to get it in there."

This story illustrates just how easy it is get bogged down in the how and not ask the bigger question of why.

#7 PRODUCTION BEFORE MARKETING. If you start producing anything — anything — before you have a clear plan for marketing, you're getting the cart before the horse. When someone tells me about some exotic thing they are going to grow, I don't want to hear about how many bushels, pounds, or gallons it will make. I want to know how they plan to sell it.

Nature is bountiful enough, forgiving enough, to reward even feeble efforts at production. But sales do not come with the rain from heaven. Markets open one at a time as we push on the door. Markets are earned, not provided on the just and unjust like the earlier and latter rain.

When you get money tied up in production without comparable marketing, you can sink faster than you can imagine. Even established businesses, huge conglomerates, wrestle with this daily. That is why you see salvage priced items, overruns and literally thousands of tons of food dumped in landfills in this country.

Production and marketing must go hand in hand. What good does it do to have 50 beautiful steers or 100 acres of corn or the prettiest crop of tomatoes in the countryside if you end up giving it all away? It's better to be content with a little production well sold than a tractor-trailer load with no place to go.

#6 LACK OF CAPITAL. This is a generic title for the idea of overborrowing and debt. Many profitable enterprises go under because their capitalization is too low for cash flow. "It takes money to make money" is true more often than not.

Of course, this is closely related to sinking all your money into land rather than information and experience. If you have land but lack the capital to plant it or stock it, you can't make any money off the investment.

While I certainly appreciate our profitable size-neutral enterprises, you must remember the context under which they remain profitable. If you overrun your headlights by jumping them up with costly infrastructure or overproduction, you can take a profitable small enterprise and turn it into an unprofitable large one. This happens all the time.

For example, anyone can have 20 laying chickens running around outside and they will scavenge all their food from nature's bounty. But when you jump that up to a 1,000-bird laying operation, with feeders, waterers, housing, walk-in refrigeration units, skids-full of collapsible egg cases and so on, you'd better have the capital to finance the expansion and carry the business for a period of time until the eggs repay the investment.

If you're going to make a mint in agri-tourism, you need the capital to get the tractors, wagons and other facilities to handle the people. All that needs to be done before the first person pays you for a visit.

Certainly I would encourage you to take the next step from the homestead chicken flock to the viable salaried enterprise, but don't overextend your capital to do it. Grow a little slower if necessary.

#5 TOO MUCH TIME SPENT IN NON-FARM OFFICES. Many new farmers, because they are not confident, spend hours down at bureaucrat offices, or talking to a banker, fertilizer salesman, feed salesman or other agri-business professional.

Everybody dispenses advice that is self-serving and before you know it the poor farmer is enslaved to the system. Many people get angry with the agri-business establishment or the bankers for this, as if it's some sort of conspiracy. But I say nobody held a gun to the head of the farmer and made him march down to take advice from these yo-yos. If these advisors and sales people deserve any blame, the farmer certainly should be held accountable for seeking their counsel. If the teacher dispenses foolishness, shame on him. If the student sits and listens to the foolishness, shame on both of them.

When thousands of farmers went out of business in the economic blood bath of the early 1980s, the conditions that precipitated the upheaval did not occur overnight. A crisis is merely the culmination of many little events. Those farmers who blamed the government lending agencies for their plight should have realized the bondage they were putting themselves in when they signed those contracts a decade or two earlier. They started down a slippery slope as soon as they went the high debt direction.

Even if you're a novice, you will be much more sensible staying at home, hoeing your own garden and contemplating quietly among the beans and corn what decisions to make. These other folks have a wonderful agenda to separate you from your money. The more you listen to their counsel and rub shoulders with them, the quicker you'll succumb to their hypnosis.

New farmers routinely jump on all the extension bus tours and get doted on by the agri-business community. Attending the fertilizer store's chicken barbecue or the chemical company's free oyster dinner may sound tempting, but it will surround you with information guaranteed to make you fail. The meal is not worth it.

When I do attend one of these extension bus tours, which is about as close as I get to these sorts of things, I invariably run into several of these newbie farmers who moved into the area with more money than brains and I just can't believe how they think. I went on a forage tour once and watched incredulously as the whole bus oohed and aahed over a "perfect kill." The demonstration was an herbicided pasture no-tilled into corn. Everybody was enamored by such effective pasture death that would allow a corn crop to be grown on a steep hillside without plowing.

I wanted to jump up and say: "Hello, is anybody home? Are we going to return this fall to see the little gullies created by denuding the hillside? Has anyone asked this farmer if he will make any money on this project?" Fortunately, I knew the questions to ask and could see the fallacy of the demonstration, but nobody else on the bus saw it. This kind of information will sink you faster than you can imagine.

#4 SPENDING AHEAD OF MANAGEMENT. Get your management down before you expand. The difference between a backyard grape arbor and a one-acre vineyard is incredible.

You can prune the backyard arbor in a couple of hours on a warm February afternoon. You can weave the trimmings into half a dozen grapevine wreaths and have the whole area cleaned up in a few minutes. Some leaves or wood chip mulch inside a wooden border makes an attractive ground cover and eliminates weeds. In the spring when buds begin to form and a late frost threatens, you can drape some old sheets over the whole arbor and protect the vines.

When summer comes and the Japanese beetles begin defoliating the vine, you can go out in the late afternoon and shake the leaves, watching the beetles roll into a pan of water you're carrying. In the morning, you can feed them to the chickens. If a drought comes, you can dip water out of a rain barrel underneath an outbuilding downspout, carry buckets of water over to the vines and keep them growing well.

At harvest, when the fruit begins to ripen, you can pick a day for juicing. In an hour, before breakfast, you can clip off all the grape bunches and the children can begin pulling them off for juicing. By noon you can have half the crop juiced and in cans and by nightfall, the last batch of grapes goes into the juicer. One long day, but the whole crop has been picked and stored. Now the arbor can wait and rest until the cycle repeats itself one warmish February afternoon a few months hence.

This all sounds wonderfully refreshing and enjoyable, and

indeed it is. But now look at the scenario when you put in an acre of vineyard. The trellising is not just a simple arbor, but a massive array of posts and high-tension wires. Keeping the grass mowed around the vines takes hours a week. Pruning takes days and the trimmings would fill a couple of dump trucks. Do you just pile it, burn it, chip it, or find someone who wants to make 1,000 grapevine wreaths?

If you're concerned about drought, perhaps you've installed a drip irrigation system. The watering system needs maintenance and looking-after. Since you can't shake all the leaves into a pan of water, you begin a foliar spray program with botanicals and soap-based stickers to deal with fungus, disease and pests. As the fruit ripens, you work dawn to dusk clipping bunches of grapes and selling them to customers who come to the farm. Other members of the family deal with customers while others continue to harvest.

Finally the last bunch is clipped off and you've been able to sell 60 percent of the crop. By manning your little juicing operation 24 hours a day for a week you were able to salvage about 20 percent of what people didn't buy, but the other 20 percent went to friends, relatives, pigs and chickens. I am not at all trying to discourage anyone from having a vineyard, but I am trying to paint a realistic difference between a "backyard anything" and a commercial enterprise.

Many farmers drown when they try to step up from a model that works great as a backyard system to a commercial enterprise. Make sure your management abilities are able to handle the jump.

#3 INABILITY TO VIEW THE FARM AS A BUSINESS. Your farm is not a romantic fantasy island any more than the city job is. Failure to put decisions through a rigorous gross margin analysis will result in many foolish decisions.

Of course, many farmers have not thought like business people for a long time. That is why they do some of the silly things they do, like borrowing $250,000 to build a concentration camp poultry house that at best may return $10,000 per year if you don't count any labor costs. That is absolute lunacy. But farmers stand in line to do it.

I know niceties and flower gardens and such are wonderful, but work on the business first. Analyze the time you're spending in terms of return per hour. Analyze your purchase in terms of return on investment. Does it pay back in one year, two years, or ten years? How about in a lifetime?

For some reason we have romanticized farming with too many John Wayne westerns and ante-bellum *Gone with the Wind* movies. That is not the real world. The real world is you making a living. Anyone not willing to subject every hobby, every nicety, every purchase and every hour of the day to a rigorous business analysis is not ready to run a successfully profitable farming enterprise.

#2 SPENDING TOO MUCH MONEY ON THINGS THAT DO NOT CREATE INCOME. Board fences are nice, but they don't make any money. Dually tonners (one-ton dual-wheel pickups) get plenty of looks at farm shows, but unless you're in the livestock hauling business putting on 50,000 miles a year, they won't pay. If you feel like you need a watchdog, you don't need a $100 purebred. A giveaway mongrel will bark just as well as a purebred.

I am constantly amazed at the things farmers buy that do not generate income. Subject each purchase to an income-generation analysis. Ask the question: "Will this make us more money?" Often what we buy is something we may want, or something the neighborhood expects, but it is not what will generate cash.

Don't spend anything unless you can figure out how it will generate more cash. If you want a sunroom on the house, don't build it until you can see how it pays for itself, either in reduced heating bills or a bedding plant trade. Maybe you can grow unseasonable vegetables. Don't even think about buying a computer until it can prove its ability to generate cash.

Until you are fully employed, hours spent on non-income activities are worthless. Saying, "It'll save some time" is not good enough if you have time to spare. And if you are watching any TV, you have time to spare that is apparently worthless. Concentrate your spending on income-generators.

#1 DISAGREEMENT OVER VISION BETWEEN HUSBAND AND WIFE. Now we're down to the number one pitfall for new farmers — disagreement over the mission between all the parties involved, especially the couple.

Iron out, beforehand, what you intend to do. Agree on a mission statement and on the action steps to get there. And then stick with it. Any deviation not agreed upon will be a major problem down the road.

Is everyone happy with a grow-your-own food policy? Is everyone happy with a one-car policy? Who is going to do the marketing? What do we really want to accomplish in this life?

These are all questions you need to address before you start the farming enterprise because major disagreements over the basics in midstream will absolutely sink the boat. Do the kids participate in little league or not? What is your policy toward pets? What is your comfort level regarding insurance?

When you do not have compatibility and like-mindedness all parties will live in a state of frustration. "Why do we always spend money on your stuff but not mine?" becomes a refrain that will wreck the business. Balance, give-and-take, a servant's heart — all these are critical to maintain equilibrium. But none of this will override fundamental disunion over the basic direction and the path to get there. It's not good enough to agree on direction and disagree on the path.

Certainly you cannot anticipate every area of disagreement, and for sure Teresa and I have had our discussions. This is the stuff of life. But basic lifestyle and mission differences are not like disagreeing over whether we should plant 20 tomatoes or 30. In that case, we'll probably plant 25.

I've watched couples move to the farm and completely flounder over this issue, and it's heartrending. If she's a cleanie and he's a messie, her required new mudroom/shower facility on the back of the house would take precious capital from the farm. If he's a cleanie and she's not necessarily a messie but maybe an outdoor type, his house-hours may come to haunt the family when the bills come rolling in. I'm not suggesting that women can't do outside work. Teresa handles the lawn, the garden and is a great tractor driver. And I certainly wash dishes from time to time. But the division of responsibilities must be agreeable to all and must accrue to an increased income.

I am not a marriage counselor or a farm consultant. All I can say is that this incompatibility over vision, in my view, is the number one way to fail. It must be dealt with beforehand, thoroughly enough to not be rehashed every time a little discussion comes along.

While I do not purport to know how many of these principles for failure must be practiced to make the whole farm enterprise fail, I can tell you that it doesn't take many. And I can guarantee you that practicing just half of them will doom any farming venture to failure.

The Ten Worst Agricultural "Opportunities"

In this chapter we are looking at centerpiece enterprises. Every farm has a centerpiece and then, if it is healthy, complementary additional or supporting enterprises. All the pieces should complement the centerpiece in a synergistic way. Keep in mind here that we are dealing strictly with centerpieces.

#1 SEEDSTOCK ANYTHING. Leave the purebred business to the independently wealthy folks with deep pockets. Certainly some folks have made a little money producing breeding stock, but only after they've been exemplary commercial farmers.

You don't start out trying to produce the best of anything. Ask yourself: "Would I buy breeding stock from me?" The whole seedstock community is like a fraternity, an exclusive club. In order to play that game, you have to look the part.

That means you need a nice gooseneck trailer pulled by an extended cab diesel dually, preferably with TV antenna on top. You need to attend the social functions, sales and meetings of the purebred organization. You need a perfectly manicured driveway lined with white designer board fence. A landscaped barn area complete with flowers and boxwoods should complement the expansive house with adjoining summer cottage for visitors.

The prestige and look associated with producing and marketing seedstock ensures that it will never pencil out. Yes, a few people have been able to sell seedstock without some of the trappings, but I don't know anyone who has been able to do it without any of the trappings. Printing slick sales brochures and keeping things impressive enough to please buyers takes a lot of time and money.

The folks who have been able to carve a niche without all these trappings have been in the regular commercial business a long time and distinguished themselves with a track record of excellence. Producing seedstock is something that either caps a distinguished career or consumes a pile of money right up front. It can only happen one of those two ways.

The press loves to publicize the expensive stud that went for thousands of dollars at the special sale. But the reason that critter is famous is because it was at the top. It also doesn't tell you whether or not that animal made a profit. Many times the animals or the plants that win the shows are judged not on function, but on artificially enhanced production capability. Corn that can scarf up more chemical fertilizers can win a trial. A bull that can convert more grain and hormones to bulk is the winner. It doesn't matter that the offspring of that bull can't reproduce. It doesn't matter that the corn is more drought-prone.

What judges look for in shows and in seedstock exhibitions often has nothing to do with functionality, but formality. And such criteria are as inappropriate in plants and animals as they are in buildings. And if they do require function, what kind of function is it? Is it hardiness without pharmaceuticals? Is it disease resistance without pesticides?

Unless a seedstock operation can be profitable on commercial prices, it is not really viable. For every animal sold successfully for breeding, a couple must be sold at regular commercial prices. Seedstock production requires a lot of hype. If you ever produce seedstock, let it be as the climax to lifelong excellence and development. Then the merits of your work can speak without all the glitz and glitter.

#2 EXOTICS. This would include fallow deer, emus, alpacas, ostrich, alligators and related animals. It would also include specialty plants like rutabagas, salsify and turnips.

The reason these things are exotic is because the market niche is small. They are not things people buy routinely. These things run on a curve like this:

· *News:* a new animal or plant gets some press. Somehow people hear about this latest greatest thing and it's purported to be incredibly lucrative. Along with this is the notion that everyone wants one, it's a producer's market, and the future is rosy.

· *Hype:* a flurry of interest rises from this publicity and everyone jumps on the bandwagon. People buy into the enterprise, paying huge prices for this exclusive breeding stock. This is the new big thing.

· *Disillusionment:* the market crashes. Suddenly all these folks that got into it look around at each other and realize a market does not really exist. It's all artificial, created by the early hype. The only reason the first folks made money was because enough people jumped on the bandwagon to create a market not for actual use, but for production only.

· *Trough:* things hit bottom. Breeding pairs that sold for thousands are now worth a couple of hundred dollars. Folks lick their wounds and go out of the business by the droves. But this brings the production numbers down to what is a realistic demand market. This is how many we need.

· *Steady growth:* slowly, but steadily, the price begins coming up but it will be years — maybe forever, before the lost capital investments get covered.

We have seen this cycle time and time again with exotics. Ostrich farming was the best thing coming a decade ago and people bought breeding pairs for $20,000 apiece. Now disillusioned producers in Texas are just opening the gates and turning them loose. They are the ultimate Texas roadrunner.

My rule of thumb for exotics is that as soon as you see the item addressed in a USDA publication, you're too late. The people who make money in exotics are the folks who get in before anyone knows about it. Once you start reading anything about it, the opportunity is already gone.

I would include in this category anything that is new and gets a lot of USDA press. For example, down in southwest Virginia a few years ago, as tobacco prices plummeted, the extension service came in with a new crop that would replace tobacco: broccoli. Much of the plant establishment work could be done with tobacco type machines and it seemed like a good fit. Nice, high value crop requiring similar soils and fertility. The only problem was all the farmers read the same bulletin. Everybody jumped on the bandwagon and ended up plowing down tractor loads of broccoli because the market had not been nailed down for this influx of new production.

A similar thing happened with shiitake mushrooms. For awhile everybody and their brother were growing these scrumptious delights. But now most of those folks are out of business and the price is high again. Now some are getting back in, but more conservatively.

The same things happened with Christmas trees. For awhile every extension forestry publication touted Christmas trees as the hottest forestry deal going. Thousands and thousands of acres later — and several years until the glut caught up — the price plummeted and folks who had planted trees 5-8 years before expecting to easily pull in $25 per tree suddenly had to accept wholesale prices at $5. Some areas are now selling trees for this price retail.

Just remember that the reason something is exotic is because it is hard to do. Buffalo are a prime example. If you saw what kind of corrals, fencing and trucks these wild dudes require, you'd think twice about getting into them. They are not easy to control, manage or produce. And they are not easy to market, either.

Yes, some folks have established a market and have carved out a niche for themselves. But carving out a profitable niche with an exotic item about which the normal person is completely unaware is extremely difficult.

I went to a conference where a fellow was hawking his emus as the latest greatest. I stood a little bit away and watched how he interacted with people. His extended cab diesel 3/4-ton pickup was parked adjacent to his exhibit area and he confidently expounded all the virtues of this exotic bird. To listen to him talk you'd think we were on the verge of McDonald's serving emu burgers!

After he spied me standing there, he came over and asked me if I could come back later to talk privately because he wanted to ask me a few things. When I returned, he confessed that his problem was marketing. "I have a whole field full of these things and can't give them away," he lamented.

I have no ill will toward the fellow. He was simply trying to create excitement for his product and get people to buy. Nothing wrong with that. But to see the difference between the salesman and the confessor was quite illustrative of the whole exotic roller coaster. Stay off it.

#3 PET LIVESTOCK. The underlying problem with these critters is that they are not consumable. How many times do families buy a pet hamster or a pet rabbit? How many times does a family buy a Vietnamese pot-bellied pig?

You have a one-shot deal. From a marketing perspective, the hardest part is always getting the customer. Once you get the customer, you want repeat business from the same person. That is how you build customer loyalty. No business can survive long if it continually has to corral brand new customers.

While it is true that a satisfied customer is your best advertisement, you are still locked in to just those people who want *any* pet, and then a smaller subset to those people who want *your* pet.

Beyond that, people who buy these critters routinely have problems. The animal gets sick, they get tired of it, or it's just not what they expected. Inevitably you are the person to unload on. Do you take it back? How much time do you devote to giving advice? How patient can you be with an ignoramus?

If you have a great pet store(s) with a price and volume contract, then go ahead and produce some of these. But speculating on pets is a no-win situation. You have to go where many people are to effectively market, or spend a pile of money advertising. That throws you into a high cost direct marketing campaign or a low-priced wholesale commodity program through a retailer. Neither one will give you much profit.

#4 HORSES. What can you do with a horse to make money? Think about it for a minute. Make a list. Okay, how many of those things are you doing or will you do? Let's look at some of the items on your list:

- *Boarding:* okay, but how many people need this service? Yes, you hear about people paying $100 a week to have their horses boarded, but I know lots of people who have gone out to find them, and they're always in the next state. Or they just died. Or they just sold their horse. This is that mystical, elusive easy money deal. If you do happen to find someone like that, you have to be an expert, provide a freshly bedded stall every day, and make sure that horse gets daily exercise. Do you know how much time that takes? Do you know what kinds of facilities horse boarders require? You could live in most of them I've seen. Anyone who does not require such meticulous care won't pay more than $50-$100 per month. But they'll still want to be able to come out and ride anytime. They still want it clean and presentable. And you still have to feed it and control it. Forget it.
- *Riding lessons:* some people are willing to pay for riding lessons, but are you qualified to instruct them? People are not stupid. They know what your abilities are. In your area, assuming you had the expertise, could you find enough customers to make it pay? Do you know how much it costs to keep a horse? Do you know how much liability insurance is for such an outfit? In order to pay the insurance, you're going to need a lot of students, and that will take lots of time.
- *Training/Breaking:* if you feel good enough to do this,

more power to you. You know much more about the opportunities than I do and will probably do okay.

· *Pulling carts or wagons:* some people rent their horse and wagon out to businesses for local parades. It adds attention-grabbing nostalgia to the publicity. Now you need a wagon, harness and a trailer to transport the horse. Yes, money can be made here, but you have to be good. I do think this is an untapped possibility.

· *Trail rides:* now you have to own several horses, or be involved enough with horse enthusiasts that they will trust you to lead them somewhere. While I understand that plenty of people do this and actually earn some money doing it, getting there is almost always a hobby, a recreational pursuit. The people who really devote spare time to this recreational venue end up becoming good enough to actually earn a few bucks. Being near a tourist area is a big advantage on this one.

· *Racing:* come on, get real. Go visit a few of the folks who earn just enough to pay the upkeep on a Kentucky Derby winner and tell me you're ready for that. If you can make that enterprise work, you don't need this book.

· *Trading:* buying and selling horses requires tremendous expertise, along with hauling and boarding capabilities. This niche is probably harder to break into than the seedstock fraternity.

The bottom line is that horses, in our culture, are recreational animals. Yes, you might pick up a few bucks with one, but it will never pay the upkeep. That's not to say horses don't have a wonderful recreational niche and some people have certainly made money with them. But they are the exception, not the rule. I like horses as long as somebody else takes care of them.

#5. CAPITAL-INTENSIVE ENTERPRISES. Anything that requires a $20,000 machine to make it work is not a good bet. For some reason we have become enamored with fancy machines and humongous structures. I know plenty of people build ponds and raise catfish, but most work in town to really pay their bills.

Sinking capital into things that rot, rust and depreciate is risky and enslaving. You have to keep cranking product through that machine or that structure just to pay for it. I know loggers who own half a million dollars' worth of machinery who don't make as much as a guy with a pickup truck and a chainsaw. The guy squiring around a knuckleboom loader, two log trucks, a skidder and a bulldozer has to keep cranking volume to maintain cash flow. If buyers put a quota on, which they do regularly in overage situations, he's got a huge payroll and debt to pay without anything coming in. He must juggle massive jobs and keep track of people, places and multiple job sites.

You would be surprised how many huge heavy metal operations are working on tiny net profits, if any at all. On average, farmers make a lower return on investment than any other group of people in society. I can't figure out why a guy would want to invest half a million dollars in something and work long, hard days, just to net out $20,000 per year. If you have that kind of money to invest you might as well put it in the bank and collect interest.

#6 CONFINEMENT ANYTHING. This is state-of-the-art agriculture, but it is the new feudal system, a self-imposed serfdom by farmers. Poultry, swine and dairy now enjoy the benefits of concentration camp production models in which the farmer borrows all the money for the privilege of allowing a vertical integrator to decide whether or not he gets any prisoners.

The poultry system is farthest along in this regressive production model. Farmers build houses that cost $250,000 once they are equipped and then the poultry company puts birds in the house. But the farmer is not guaranteed any flock beyond the one that is already in the house. The company can refuse to put birds in there for any reason.

Hundreds if not thousands of producers have lost or nearly lost their farms when the company decided to quit putting birds in the houses. The mortgage payments continue whether any birds are in there or not. The giveaway to me that these things are a dead end is that most farmers who have these houses still work off the farm for income. If these houses actually paid anything, why are the owners still working in town? How many other people do you know who would invest $250,000 in a business that did not provide half a salary?

Contract growers must be the unhappiest group of farmers I've ever met. They are even more unhappy than cash grain farmers and I think it is because they have completely lost control over the decision-making on their farm. Confinement dairy farmers are crying as milk prices continue to plummet to world market levels. The cows suffer all sorts of physical ailments, refuse to breed, and quickly go out for hamburger.

Indeed, confinement anything is one of the worst agriculture opportunities.

#7 MONOCULTURE OR SINGLE INCOME ANYTHING. You can usually tell the farmer that's bought into this poor opportunity because he identifies himself with his particular commodity:

- "I'm a cattle farmer."
- "I grow sweet corn."
- "I raise catfish."
- "I'm a goat rancher."

Farmers who identify themselves like this are throwing all their eggs in one basket, not only in production but also in marketing. Usually they have only a couple of markets and often only one. They have very little say about pricing or quantities.

This type of agriculture is strictly production oriented, commodity based, and subject to the huge fluctuations in wholesale prices. Wholesale prices are kind of like the end of the whip. Did you ever go sledding down a hill with a bunch of people forming a train? You know how the front guy could weave just a little but it made the rear guy fling way around?

That's the way it is with wholesale and retail prices. Retail prices are like the guy in the front of the sled train. Farmers are like the guy out on the end of that train. Just a little bobble in the retail sector produces a huge whipping effect in the production sector. One reason is because the final consumer is the most price-conscious. Just a few cents up or down create huge variations in purchasing decisions. At the retail level, nobody just *has* to have that T-bone or that fresh head of lettuce.

How many of us won't buy certain things unless the price is under a certain level? We can take it or leave it. For a couple more cents a pound, we'll just eat something else that week. The farmer, on the other hand, has invested huge amounts of capital to produce this product. He has tooled up, trained, and borrowed money for this product. To make the farmer quit or change production plans requires a major shift in price, often as much as 90 percent.

Of course, this is what farmers constantly complain about, but it is necessary to make the supply and demand remain anywhere close to equilibrium. Putting all your capital, time and energy into just one commodity, as a general rule, is one of the most risky things you can do in farming.

In order to produce enough of this one item to make a living, you'll have to do it big time, and doing it big time throws you into the wholesale market and all the other negatives involving huge enterprises. These include being subject to weather, pestilence and price on the production and marketing end. Recent data compiled by agricultural economists show that in order to produce a $15,000 salary, a Midwestern grain farmer must grow 1,000 acres of corn or 3,000 acres of wheat. Harvesting alone is a major nightmare both economically and logistically. That is not a realistic option for a would-be farmer. In fact, if the truth were told, it's not a realistic option for most of the farmers who are doing it.

The bottom line is this: stay away from monocultures and single income commodities.

#8 ANYTHING YOU DON'T EAT ROUTINELY.

If you don't eat it frequently, chances are nobody else eats it frequently. Keep in mind, now, that we are dealing in this chapter with centerpiece enterprises. Plenty of opportunity exists to surround a common food with some uncommon ones. That's perfectly sensible.

But exotic, strange, seldom-eaten products are not the way to bet for your main enterprise. What you want for the backbone of your farm is something folks are familiar with and consume in serious quantities, except you want to be able to produce that item enough above normal to carve out distinctiveness in the marketplace.

You can look at other items we've already listed, like some of the exotics, and realize that this principle gives it two strikes. In fact, the more of these principles the item violates, the worse it is. Stay with the familiar and shun the weird.

#9 ANYTHING YOU DON'T REALLY ENJOY. This would include anything you're only doing for the money. Success takes too much effort to try to attain it as a mercenary. Money alone will not sustain the commitment and passion required to be successful.

I've spent a lot of time in this book trying to help you be reasonable in your approach, to think logically, to act like a business person. But here is a vote for your heart and your intuition. You need a balance. I confess that I really don't enjoy being underneath horses. Chances are I shouldn't do that.

I really don't enjoy doing mechanic work or meticulous carpentry work. I'm not good at it and I just don't like it. These are part of who I am. Chances are I should not be a cabinetmaker or an automobile repairman. If you don't like animals, you probably should not do a livestock enterprise. Or if you feel lovey-dovey about all animals, you probably should not raise them either — at least not for human consumption.

If you don't like plants, you probably shouldn't be a truck farmer. Be assured that no matter how much you enjoy an enterprise, you will find parts of it that you dislike. I'm not talking about those little pieces. This deals with your soul, the inner you, your temperament and what turns you on — or off.

Something that you've never been interested in probably should not be the centerpiece of your farm. When magazines come, note which articles really fire you up and which ones you kind of skim over. You might want to make a list, just to help you discover who you are. A pattern will develop and over time you will discover something: "Hey, these articles consistently get read but these consistently get overlooked."

Then ask yourself for the common thread. Are the articles you're constantly reading about animals or plants? Or is it something a little finer, like food or ornamental; perhaps pet or gourmet food? If you're afraid your natural inclinations are leading you in directions that will make farming extremely difficult — which violate several of these principles — then perhaps you should visit some farms that practice the things you find disdainful and see if your antipathy is well-founded.

This is close to my heart because we do a lot of animal processing here — also known as butchering. Many times people will come to experience it in order to find out if their intuitive uneasiness is real or founded on ignorance and misconceptions. Most of the time people come away with a completely different outlook.

I think this caution is in order because our premonitions about things often have no basis in fact, but rather are subconscious reactions to prior events or even stories. Anyone who has gone through the child-naming process should be able to appreciate this. When you were picking names for your children, were there some that you just couldn't stand? Generally, the names we really dislike are names of people who left bad impressions on us. Maybe it was the name of the class bully. Or maybe it is the name of the girl that always beat you out for the lead in the school play.

The same is true of names we like. Sometimes we can't understand why we like a certain name until we remember, "That was the name of that sweet high school girl who sat next to me on the bus when I was in second grade and enjoyed letting me read to her." Or maybe that name you like belonged to the hero professional football player that your older brother idolized when you were kids, and big brother always made good choices.

You see what I'm saying, I think, so just don't run with this one clear off the end of the balance. All things being equal and making decisions from a point of understanding, stay away from things you really don't enjoy.

#10 ANYTHING YOU THINK IS A WAY TO GET RICH QUICK. If someone told you about a farming enterprise in which you could get rich quick, don't do it. I would include here even subconscious thinking. If you even begin to think that such and such a centerpiece enterprise is easy money, fast money, or both, forget it.

"There's no free lunch." It's true. Oh, it's true. I can think of only two reasons why someone would try to sucker you into a deal with this promise. Either they tried and failed so miserably that they want to drag as many other people down into it as possible so they won't be the only ones foolish enough to do it, or they're playing fast and loose with the truth.

Perhaps they haven't finished paying all their bills. Perhaps misery loves company. Perhaps the deal hasn't gone full circle yet. Or perhaps they are just naïve. Whatever the reason someone might try to convince you otherwise, forget the easy money scheme. It does not exist.

To be sure, if I didn't think sound, profitable farming enterprises existed I would certainly never have written this book or the other two. Be assured that this book is not being written as a way to get rich, either. It is in direct response to the countless people who ask: "How can I do what you're doing? Where do I start?"

I certainly hope I succeed not only in honestly portraying the opportunities in farming, but also the necessary discipline, perseverance and sacrifice required to get there. If I knew of any shortcut, I'd certainly take one. But it's not there. Period. Generally, every market opportunity that sounds too good to be true, is. Anyone who says they will take all your produce, won't. Anyone who has found a way to legally get rich quick, hasn't.

Well, there they are — the ten worst farm centerpiece enterprises. To be sure, a few folks, who were in the right place at the right time, have made successes of each of these. I am not suggesting that it absolutely cannot be done. But if I were a gambling man,

I wouldn't put money on any of these enterprises, especially if I were a new or beginning farmer.

Stay with what makes sense and steer away from anything that smacks of hype as the primary way you will make your money on the farm. As your farm becomes successful, you can always tuck a couple of oddball projects around the edges if you want to. But to plan on these as your farm's backbone will spell disaster.

Best Centerpiece Agriculture Opportunities

While no one has comprehensive knowledge about all the viable entrepreneurial agriculture activities, certainly some deserve specific mention and an explanation as to why these are good opportunities. The point of this discussion is not to channel everyone into certain enterprises and then cause a glut. The goal here is to publicize, right now, some of the most viable farming enterprises.

I've used several criteria to make these picks:

- Low initial start-up cost relative to the ability to generate income. Generally the lower capital-to-annual gross sales ratio is best. It doesn't mean no capital costs exist; it just means that relatively speaking, a high income generation potential can be accrued from a relatively small capital investment.

- High gross profit margin — a big difference between expenses and income.

- Relatively low maintenance requirements. This includes minimal depreciation, high customer return rate (highly consumable product) and low capital replacement costs.

- High cash flow relative to expenses so that it generates free cash. Included here would be a relatively quick turnaround time between investment and payback.

- History of high success rate among new enterprises.

- High demand, low supply in the current marketplace. Plenty of room for expansion — in most cases, quadrupling would be simple. In conjunction with this, none of these is currently encouraged by USDA, meaning mainstream folks have no idea about these potential moneymakers. Unless this book sells a million and every reader does one of these enterprises, there is plenty of room for you.

- High product distinctiveness, making niche marketing easier.

- Relatively size-neutral profit potential. Profit margins do not decrease appreciably in smaller enterprises, making it easier to "get in" with prototypes and still make a profit.

Remember here that we are talking about centerpiece enterprises. Also remember that each section of the country is different and will favor one more than others. Where this is especially applicable, I will try to point it out.

#1 PASTURED BROILERS. In all honesty, I think this is probably the most golden agriculture opportunity for the foreseeable future. Here is why:

- My first book, *Pastured Poultry Profits: Net $25,000 in 6 Months on 20 Acres*, details this prototype in a way that makes it easy to duplicate anywhere in the country. You do not need to reinvent the wheel on this one. It simply works.
- Everybody is eating chicken. Per capita consumption is now higher than beef and trending up. It is highly consumable —

Photo 10-1. *Pastured broilers start out in a brooder house on deep bedding before going to pasture. Small children enjoy taking care of the chicks at this stage.*

Photo 10-2. *At three weeks old, the chicks go out on pasture in floorless 240 sq. ft. pens that we move daily to a fresh salad bar. No manure to haul, to smell, wonderful fertilizer, easy marketability, and tremendous profit potential make this one of the most profound opportunities of our era.*

people eat it almost every day.

• Extremely low start-up costs. You can start this in your backyard. People have started with a couple of pens in the middle of the city or in a garage.

• Highly portable — if you change locations, the entire infrastructure can be moved on a pickup truck.

• Quick cash turnaround. From first chick to finished broiler is 8 weeks. You have your money back in 8 weeks! That's almost as fast as radishes.

• Pastured broilers taste so much better than anything else — including loose-housed, range-yarded organic — that they just sell themselves. Of the countless people who have tried this enterprise, I've only encountered a couple that mentioned having trouble marketing. For most, this is the biggest fear going in but turns out to be the easiest part of the whole shebang.

• Complements almost any other enterprise. This to me is the real plum. It does not compete with other land uses. It can be done in an orchard, around the mowed edges of a vegetable farm, in a field of cattle or sheep, and in the paddock where horses exercise. It can be done along the grassy edges of a lane or in the median of a highway. The expansive lawn of a country club or industrial facility can be fertilized and maintained with pastured broilers. That idea might take some real slick sales pitching on your part, but once a few people — maybe you — break the ice, it will become the fad.

• Relatively free market access. Government regulations allow producer-growers to process and market their birds to individuals and the hotel, restaurant and institution trade (HRI). More than 30 states allow up to 20,000 birds annually, which is the federal exemption from inspection. Some states are quite a bit more restrictive, some down to 1,000. But creative structuring from a business or marketing angle and a little bit of pioneer savvy can keep the government officials scrambling.

• Support network exists. The American Pastured Poultry Producers' Association specifically targets practitioners, providing a catalyst for creativity, problem solving and information sharing.

Photo 10-3. *The farm of the 21st Century—clean, green, humane. Highly portable, low capital and high return.*

• Child-friendly. All children like little chicks, and this enterprise concentrates on young birds. The chickens will not kick a child in the shin or run over him in a corral. A child-sized critter is fun and one that the whole family can enjoy.

As the industrial poultry model gets more unnatural, the niche for pastured chickens continues to widen. Because the commercial media expose new food and employee abuses in the poultry industry almost every day, consumer desire for a change is incredibly high. To find folks who want to opt out of the conventional fare is easy when *Time* magazine reports that roughly 10 percent of the weight of supermarket chicken is fecal soup. People want to know where they can get something different.

From Colorado to Alberta, and Oregon to Ontario, Belize to Saskatchewan, people are running pastured broilers with great success. If I knew of a better opportunity, I'd be doing it.

#2 EGGS. I would like to say pastured eggs, but that is not completely necessary. Layers are more adaptable to different production models than broilers, which are young and lethargic. But egg layers are mobile, much smarter, hardy and more aggressive in what they will eat.

Eggs are less regulated than broilers, and certainly just as consumable. The difference between battery-raised eggs and anything else is incredible. And especially when people hear a brief description of concentration camp egg production, you will find them wanting something else just for humane animal production reasons.

The industrial model sports a house as long as a football field with three tiers of cages. Each cage is about 22" by 16" and houses nine birds. In most, at least one dead bird is in some stage of decomposition, being pushed through the mesh floor by its former cellmates, who hardly have enough room to squat. The house stinks to high heavens, is dusty and dark, and generally looks like the bowels of the catacombs rather than a light, airy birdhouse. A steady diet of hormones, medications, cooked chicken feathers and guts, as well as grains is touted as a "balanced, natural diet." Without a nest, the birds drop their eggs on the slanted wire mesh bottom of the cage. The eggs roll to an outside edge where they begin a long journey via canvas conveyor belt to the processing room.

Adult layers can consume prodigious amounts of grass clippings if you do not have enough room for a pastured operation. In addition, they love sprouted grains, hay chaff and other pasture-type things. They will even eat huge garden weeds down to the stem.

Because of their mobility, layers can be incorporated in many ways within existing enterprises. Free range chickens from portable egg houses — see *Pastured Poultry Profits* — can sanitize under orchards or rabbits. Any feedlot would support a sizeable adjacent egg laying operation — the layers free range through the cattle, picking out undigested grain and eating any insect larvae. The same would be true for the lounging paddock for a normal dairy — every cow will pass enough grain to feed 2-4 chickens, depending on volume of grain used. A horse boarding facility could feed plenty of chickens.

Photo 10-4. *Pastured eggs can be produced in portable field pens virtually anywhere. Mature hens are much hardier than broilers and eggs enjoy substantial freedom from government regulations.*

Photo 10-5. *Polyface apprentices Micha and Jasen gather eggs from pastured layers during evening "egging."*

Photo 10-6. *Jason picks eggs out of the next boxes in the back of the layer pens. These eggs are good enough to command triple commercial prices, especially from gourmet restaurants.*

Egg quality is highly variable. By utilizing several more natural production concepts, you can have a highly marketable egg; it will taste better, handle better and look better.

#3 SALAD BAR BEEF. I am talking here about retail beef. The easiest way to do this is to buy heavy calves (500-600 pounds) and keep them for a season before beefing them in the fall. Although we keep cows, I would encourage you to stay away from cows in the early years of your new farm. Cows do have some advantages over calves because of their lower nutritional requirements. Furthermore, we do not depend on beef as our centerpiece enterprise, which gives us a little more latitude in determining what kind of cattle we will run.

We do, however, augment our own calf supply by buying calves from others — even through the stockyard — and adding them to our beeves. We keep only enough cows to eat junk hay and to get full utilization from pastures. If we have more market for beef than

Photo 10-7. *Salad bar beef lounging after filling up with greens. No feedlots or center pivot irrigation systems here. Herbivores build fertile soils and lush vegetation; they are still the most efficient way to do it, but we must mimic domestically the principles exhibited in the wild.*

Photo 10-8. *Micha Hamersky, a Polyface apprentice from Austria, rolls up portable electric fence: the steering wheel, accelerator and brake on the flour-legged mower.*

the calves these cows produce, we buy calves rather than increase cow numbers.

The biggest advantage in this enterprise is the minimal infrastructure: some water hose and electric fence. The only big expense is buying the cattle. But since you don't need costly buildings and machinery, as soon as you sell the beef in the fall you have your up front money back plus a profit.

Since this does not require expensive infrastructure, its profit potential is size-neutral. You can run two calves in the two-acre lot out back — assuredly they will only get 1/20th of an acre per day in a paddock, but using electric fencing makes this simple and cheap. My book, *Salad Bar Beef*, explains all this in great detail.

The market is excellent for clean, lean beef, especially if it demonstrably heals the land like the buffalo stimulated prairie grasses. Mad cow disease now stalks the subconscious of consumers everywhere and you can capitalize on this awareness with an alternative beef product. Probably 99 percent of all beef producers in the U.S. have facilities and equipment that are far too elaborate. When you take out the equipment, the expensive fencing, the silos, grain feeding and trucking, all you have is cattle, grass, a little strand of wire and some black plastic pipe.

People still like beef. They really do. All you have to do is provide delicious beef that comes without a guilt complex or fear, and people will buy your product.

#4 GRASS-BASED DAIRY. Especially in states that allow the sale of unpasteurized milk directly from the farm, this is an incredible opportunity. The only thing that keeps this from being a better opportunity than pastured broilers is government regulations. Without them, be assured that I would be doing this too.

A seasonally milked Jersey cow will produce at least 800 gallons of milk per year on average. At a retail price of $3 per gallon, that's $2,400 per cow. With good pasture (to milk, you need excellent pasture) you can run roughly one cow per acre, but for the sake of extremely conservative estimates, let's drop that to 1½ acres per cow. At retail prices, milking 10 cows on 15 acres could yield

$24,000 in gross annual income. That is downright lucrative.

Regulations notwithstanding, unconventional dairying still qualifies as a centerpiece enterprise. If you can work around the regulations, or if the regulations are agreeable in your state, so much the better.

Right now one of the advantages to this enterprise is the price supports, or dairy subsidies, as they are often called. Although they are gradually being phased out, it will be some time before a true free market exists in the dairy industry. As an unconventional producer, you can capitalize on the government-created artificialities because you are running on a completely different economic matrix than the normal confinement dairy. And the industrial confinement dairy provides the basis for mainline dairy economics.

This enterprise includes, of course, value adding cheese, butter, yogurt and ice cream. If you tool up with stainless steel equipment and processing rooms for one item, it would only be reasonable to add the others and diversify your portfolio. Until then, some areas with a prolific dairy industry have custom cheese makers with whom you can subcontract. You would want to find one that will not commingle your milk with everyone else's.

The market for organic milk is burgeoning right now with consumer concern regarding bovine growth hormone and lack of labeling. While investing in a dairy parlor is a considerable expense, many farmers have found wonderful low-cost ways to meet inspection requirements. A little sleuthing in the grass-dairy literature will acquaint you with all sorts of streamlined designs.

This differs from a beef operation in several respects:

- It requires lactating cows. It cannot be built around nonreproductive animals.
- It requires breeding. This means bulls or artificial insemination and all the scheduling and watching that breeding requires.
- You need an excellent working knowledge of pasture quality, rotations and overall grass and fertility management. Dairy cows are less forgiving than beef cows.

- Calves are part of the picture. This means both calving with its inherent birthing problems that can complicate things, as well as what to do with calves when they are born. Do you raise them, sell them, or what? Perhaps rose veal is an option.
- More consistent cash flow. While beef can be harvested effectively on grass for only about 3 months of the year, a seasonal dairy will produce cash flow for 8-9 months.
- No compensatory gain. With beef cattle, during drought or other seasonal tight feed situations, the animal can gain in frame size only. Then, when grass is lush again, they will compensate by putting on flesh in huge pounds of gain per day. A dairy animal cannot do that. If you miss some potential milk production one day due to improper management or lower-than-optimum feeding, the cow will never compensate with increased production later. Missed production is missed for good.
- Cost of the parlor. This can run anywhere from $5,000 to $100,000 depending on how many cows you want to milk, how creative your design, and how much stuff you can scrounge cheap.
- Twice (usually) daily milking. This ties you down, but no more than a pastured broiler operation.
- More experience necessary. Not only are you talking about managing a group of herbivores, but you are also talking about breeding and milking, which are big learning activities in and of themselves. I would encourage anyone interested in pursuing this to spend a few months living and working on a grass dairy first.

I've had the privilege of knowing several outstanding grass-based dairy farmers. Certainly Cyd Bickford's Wisconsin dairy is outstanding, as her husband, Paul, is quick to point out. Stacey Hall and Bill Dix in the southern Ohio mountains have carved out a fine living with a small dairy. They've grown a little at a time, scrounged used equipment, and maintained their zest for the work.

But probably Carol Ekarios in Minnesota has one of the most outstanding dairies I've visited. Without a farming background, she and her husband took some savings and bought a small farm. Then they purchased baby calves, bottle feeding them. With only $3,000 they converted 20 feet of one end of an old gambrel-roofed barn into a three-stanchion milking parlor with adjacent milk storage room. When I saw the setup I couldn't believe it was grade A. Carol explained it this way: "I got a copy of the regulations and scoured them. When the inspector came and wanted a change, I'd tell her the code didn't require that. I knew what was in there better than she did. You have to know the regulations." As a result of doing her homework, Carol was able to get by with one milking claw, a stainless steel bucket, and a 500-gallon bulk tank.

She only milks twice a day for four months and milks once a day for an additional month, yet her net return per cow is higher than the state average. And the assortment of cows is quite amazing. They are the most motley-looking bunch of cows I've ever seen. Some have a fair amount of beef in them. And to top it off, she was breeding these mongrels with a Scottish Highlander bull!

The bull was about half the size of a couple of the big Holstein cows and could not reach them for breeding. The cows could lounge in the old barn and there was a step-down from the barn to the outside. That bull learned that if he would wait inside the barn right beside that step-down, he could jump on those big cows and breed them as they exited. Listening to Carol describe how this little scheming bull did his job just about made tears flow from laughing.

In addition to the $1,000-per-cow net return, she had meat to sell from bull calves and a small flock of sheep. She contracted out all her haymaking and had six months of vacation every year. What a life!

Dairy products account for a huge percentage of the retail food dollar. Capturing some of these dollars through a low-cost dairy is certainly a viable centerpiece.

#5 MARKET GARDEN. What advice could I add to the wisdom of Eliot Coleman, Andy Lee and Eric Gibson and Jeff Ishee, who have written eloquently and prolifically? I do think your location is a key component of this enterprise as a centerpiece. The closer to a metropolitan area, the better. In fact, some of the most successful ones have been on 1-5 acres right in a city.

Successful practitioners demonstrate the viability of a market garden for generating a full-time income. Often this can be done on land that you do not own, which takes the big capital cost out of it. Here again dovetailing this operation on an existing land base can be quite do-able. Any kind of dairy or beef cattle enterprise should welcome your setting up a couple of acres in vegetables and small fruits in exchange for all the household vegetables. In fact, if you could link up with a clean beef or grass dairy or pastured poultry producer, you could get additional mileage out of the customers. If the landowner does not direct market beef or dairy products, perhaps you could take that on for a commission since you would be direct marketing your vegetables anyway.

Because vegetables and fruits return such a high gross dollar amount per acre, you can pay five or ten times the going pasture or cropland rent. This is the plum for the farmer. Certainly you could offer some labor or any other expertise as a barter or to sweeten the pot. Cattle can be fenced out easily with good electric fencing. In a garden situation, I would certainly use more than one strand — at least two, and three would not be a bad idea. Little battery-powered chargers are quite powerful and can keep you from having to plug into power anywhere. This offers additional location flexibility.

This is another wonderful enterprise to grow into. You can begin by tilling up your lawn, either by hand or mechanically. Up-front costs are extremely low. Of course, this is probably the enterprise in this list that is closest to meeting current market demand. The reason is because it is the least government regulated -- in many places you can sell vegetables from the trunk of your car at a wide shoulder on the road. Following advice on season extending, you can almost push cash flow to the whole year, especially if you add a high quality root cellar for potatoes, cabbage, sweet potatoes, carrots, squash and turnips. Good keepers can give you a marketable item almost until the first vegetables are ready the following spring.

The biggest drawback to this centerpiece enterprise, as I see it, is the relative low dollar volume per customer. The vegetable part of the retail dollar pie is much smaller than that of animal products, which means you simply do not have as many dollars to chase.

Photo 10-9. *Jeff Ishee checks his fledgling apple trees along the edge of his market garden. Rows of asparagus and garlic with portable chicken pens in the back garden areas shows landscaping intensity under profitable market gardening models.*

In order to move $30,000 worth of stuff, you need a lot of pounds of stuff, and you need a lot of customers. If the average person spends $600 per year on fresh vegetables (which I'm sure is a high estimate), you would need 500 customers in order to gross $30,000. Because the price per pound and average purchase is higher for animal proteins, we here at Polyface can do that volume with fewer than 100 customers, on average. That's a hefty difference.

One big advantage of a garden over a livestock enterprise is that the plants do not need to be controlled. They don't escape to the neighbor's iris beds or vegetable garden. They also do not need such scheduled daily care. Although timeliness is everything, as it is in any enterprise, you can set the work schedule that suits you rather than what is necessary to the animals. You can't gather eggs at 6 a.m. because the chickens haven't layed them yet. Gardening does offer more flexibility to the workday routine.

Plenty of successful vegetable operations exist and consumers crave high quality, local fresh produce. That is still by far and away the highest dollar volume seller at farmers' markets throughout the country. Jeff Ishee's *Dynamic Farmers' Marketing is* a great source of information if you want to pursue this venture.

#6 HOME BAKERY. Although you may wonder why I would put this in as a centerpiece agricultural enterprise, I think it can be the backbone of adding value to several things the farm produces. This actually allows you to grow some grain and make money at it. But being a cash grain farmer, in my view, is not a viable centerpiece enterprise. Growing a couple acres of specialty corn, however, and making home baked cornbread muffins, is a viable centerpiece enterprise. I think people who enjoy baking of any kind, and especially if they have a flair for marketing, can make a real go of this enterprise.

The primary investment is labor. Yes, a federally inspected kitchen may cost $10,000, but that's no more than the cost of a used car. The value of material you can put through there in a year, as a percentage of the capitalization cost, is truly remarkable. Specialty baked items are in high demand. Perhaps you and a couple of friends could pool your resources and build a shared facility. The market for high quality meat pot pies is not being touched. Food trends indicate that the market for prepared foods is continuing to escalate, but consumers want higher quality. They want designer food, food with character and a story — especially clean, local, fresh food.

If you use your own eggs, grain and seasonings, you would be surprised what a half-acre of barley or amaranth could be worth. Cottage industry-sized grain mills work well. Italy is the world's leader in small-sized field equipment, including the BCS line of garden equipment.

In most farming areas you can get access to a combine. If you cut the grain stalks and pile them on a pickup, you can run it through a farmer's combine for a few dollars and take home your sacks of grain. Document all this with some high quality 5"x7" color prints that you can put in a scrapbook out at the farmers' market booth or in your home sales area. Consumers love to think that their food is part of a story, and telling them your story, or the story of what they are buying, will enhance your image of distinctiveness and nostalgia.

A bakery has the advantage of allowing you to produce multiple items from a single product. You can make rye bread, rye rolls, rye crackers, whatever. You can diversify your market portfolio just by using different recipes rather than by having to grow whole different varieties of produce or animals.

If you do not want to direct market these items, certainly local restaurants, bakery shops or coffee shops look for distinctive products to enhance their offerings. This enterprise is ideal for large families, where children can be an integral part of the production. Incorporating a home bakery as part of the homeschool is wonderful. Children learn to measure, to multiply, divide, add, subtract, read recipes and learn chemistry.

I'll mention one other little asset that the bakery enjoys: mail order possibilities. With overnight package delivery, you can actually send things across the country in less than 24 hours. While this may be a small niche, you would be surprised at the number of hits some people are getting on the Internet for specialty food items.

What is the package and delivery cost if you're sending something for a gift? So what if a dozen muffins are $20 instead of $10? When it's a recreational/entertainment purchase and all you have to do is pick up the phone to take care of it, our hurried, harried world thinks $20 is a bargain for the convenience.

Many of the $1 million-plus-supermarket bakeries got their start as cottage businesses. I wouldn't suggest that you make that your goal, but I do believe that with each of these superbakeries come new opportunities for home-based, high quality enterprises. And if you add the production phase, so that you are actually turning your own farm products into ready-to-eat fare, you have an edge on the competition.

#7 BANDSAW MILL. Tree farming, in my view, is only viable if you do not sell wholesale. Most people with a bandsaw mill actually make more money custom milling for friends and neighbors than they do milling their own logs.

Photo 10-10. *Utility logs cut from your woodlot or brought to you by a neighbor can be turned into gold on a bandsaw mill.*

Photo 10-11. *Just a few logs can be turned into hundreds of boards on a bandsaw mill. Many models exist, but they all use the same principle: the log stays still while the bandsaw moves through it.*

Photo 10-12. *Stickering, sorting, and storing fresh-cut lumber under roof allows you to have a poor-boy lumber store in your backyard.*

Photo 10-13. *Storing lumber in rough-made sheds is cheap and adds value as the boards air-dry.*

The lumber industry, in case you haven't noticed, has become extremely centralized. The only reason this has not been reported in the press as much as the centralization in the poultry, beef and swine industry is because trees don't bleed. But similar environmental degradation occurs in forestry as a result of centralized, mega-processing facilities.

A couple of decades ago our county had more than a dozen sawmills, not counting the many little mills on individual farms. Today we have only three, and they buy logs from a huge area. Most adjoining counties do not even have one anymore.

An intense need still exists for someone to get a log milled into boards. Many hobby woodworkers and even more commercial fine woodworkers crave a source for good local lumber, especially unusual woods.

Right now the average price difference between stumpage and retail is 1,000 percent per board foot. That means the average difference between what the farmer gets for his wood and what he pays for it at the hardware store is the difference between 30 cents and $3 per board foot. Either way, plenty of room exists for small operators to not only provide a custom service to farmers and woodworkers, but also to turn the trees in their own woodlot into valuable lumber.

To be sure, much of the lumber is just grade material, or utility, and does not command real high prices. But at the same time, many projects can be done with lower priced lumber. For example, if you sell a walnut tree on the stump, the sawmill buyer will discount it for every little defect. A catface on the log will drop the value as much as 25 percent.

But to a woodworker, those catfaces, or burls, offer a unique quality to a board. They add personality and character to a table or hutch — they'll pay extra for the very board the sawmill discounted. That's just the way the industry works. This industrialization in the lumber industry, this movement toward sameness, creates huge marketing opportunities for unique boards. Every time I go to a craft fair or woodworking exhibition, I hear artists complain that they cannot get good lumber with distinctive character.

Anyone who has had the privilege of listening to Paul Easley from Illinois talk about his bandsaw lumber business knows that this opportunity is real. He began milling lumber from his own small farm forest and within six years what had begun as a way to earn some extra money on the side grew into a full-time business. He and his wife now make what many would consider a lucrative salary from their bandsaw operation. He saws stock material for hobby woodworkers.

For example, he sells thousands of wooden ballpoint pen cartridges: 35 cents apiece and three for $1. With a twinkle in his eye, he quips: "You know how many half-inch by half-inch by 6-inch blocks I can get from a 12" X 12" piece of black walnut?" In the early days, he started a couple of woodworking clubs in his area in order to stimulate demand for his specialty wood.

He cuts off catfaces and interesting knobs, puts them on a shelf as specialty items, with a price tag of $5-$10 apiece. "They move," he says, matter-of-factly. Now his shop is a major dumping ground for tree surgeons in the area, who do not want to pay expensive tipping fees at the landfill. Many of these trees have knobs and character that simply give his work more distinction. He takes logs that we would cut up into firewood and, with an eye for artistry, cuts that piece into $500 worth of lumber — all in a couple of hours.

It just goes to show that there is always room at the top – always room for a new idea. Always a new market to be created, or a new niche to fill. We can never saturate opportunity because the human spirit is too creative.

In many areas, bandsaw operators stay as busy as they want to be just custom milling logs for people. Centralized milling creates mega-sized logging operations that do not want to mess with small volumes of timber. Trying to get a logger interested in a 2-acre harvest is almost impossible.

This size discrimination, along with the farmette revolution, creates plenty of opportunity for folks with a farm tractor equipped with loading forks, a trailer and a portable bandsaw mill. Every single bandsaw mill operator I have talked with says the same thing: "I can't begin to keep up with the demand." Most need to be booked six months in advance if you want their services.

I think this is a golden opportunity in agriculture.

#8 SMALL FRUITS WITH EMPHASIS ON U-PICK. Highly perishable small fruits always command a premium price. Shipping and handling simply destroy these tender fruits. The difference between mass-produced, packaged, transported and stored small fruits and the ones you get just a few hours old at the farmers' market or from the neighbor is just beyond comment.

These include primarily strawberries, raspberries, blackberries, and blueberries, but more exotic specialties like currants, juneberries, and gooseberries can be grown as well. I would include in this enterprise seedless table grapes, simply because the demand is high, they are perennial and the selling price is attractive. In addition, they have reasonably good shelf life and enough varieties exist that you can offer types people in your area can't get anywhere else.

Certainly any centerpiece operation would include some fruit trees and maybe some nut trees just to round out the portfolio. Spaced widely, this can allow you to grow two tiers of fruit on the same land area.

The biggest drawback of small fruits is perishability – that is why they command a high price. They are also vulnerable to unfavorable weather conditions like cool and wet periods which are conducive to fungus growth, but that is why you want to have a few of each kind since they mature at different times of the year. If one crop fails, chances are a couple others will be good.

Diversity also will space out your cash flow. All of these have relatively narrow harvest windows, but if you have multiple types, you can spread the harvest throughout the season. Otherwise enough of any one type to make a living on will require expensive labor and machinery to get harvested in a timely way.

I am always amazed at the way raspberries and blackberries sell at the farmers' market. People who want these do not seem to care about the price. Still, these berries are a little like filet mignon — the buyers don't care about price, but not too many people are buyers. It's one of those things that if you have to ask how much filet

Photo 10-14. *Grapes can provide good supplemental income, especially if you have a customer base.*

mignon is, you probably can't afford it.

Plenty of people have built successful U-pick enterprises using strawberries as the centerpiece. This crop requires a high level of horticultural skill and management, but the potential return per acre is huge. A U-pick operation has the advantage of adding the recreation and country experience to the food, enabling you to sell more than food. Any time you can create an experience without adding appreciably to the food cost, you will find people ready to patronize your product.

You may be surprised by my top picks, and certainly everyone is entitled to an opinion. Plenty of diversity exists within this list to offer something for both plant and animal lovers, as well as artistic craftsmen types. I could have included bedding plants or catfish, but to make a full living off the net profit of these enterprises, I think, puts them too far into the capital-intensive model.

More than anything, I want you to be successful. And in my humble opinion, I think these enterprises are the most do-able for the most people.

Photo 10-15. *Neighbor and friend, Jeff Ishee, author of* Dynamic Farmers' Marketing, *checks his raspberry blooms early in the spring. White Dutch clover planted between the rows provides a perfect alley to run his Andy Lee-styled chicken tractors. A more intensive and complementary approach to land use would be hard to find.*

The Ten Best Complementary Enterprises

Once you've started in on your centerpiece enterprise you will want to hang other things around it to make a more stable whole. Generally these complementary elements should be added before the centerpiece generates your full-time income. In other words, do not wait until the centerpiece is making your living before you begin developing additional sources of income.

Certainly the criteria for these are similar to the centerpiece ones, but with the following additions:

√ *Additional use of existing infrastructure.* Concentrate on things that will better utilize space in buildings or the number of jobs performed by machinery.

√ *Filling low work times with cash generating jobs.* In the normal ebb and flow of any vocation, we want to be careful about adding additional work requirements during peak periods. We only want to add enterprises to the work valleys. Accountants, for example, try to pick up tax planning work after April 15 rather than trying to add new clients from January 1-April 15.

√ *Evenly distribute cash flow.* Just like the time factor, all businesses experience income distribution peaks and valleys. Think about an enterprise that generates income during the off season.

√ *Spreading customer purchases.* If 90 percent of your customers are vegetarians, adding animal protein to your farm may not be wise. On the other hand, if 90 percent of your customers are gardeners, vegetables will be a hard sell. Think about compatible items.

√ *Increased customer flow.* If you already have good customer flow to your sales area during the summer, try to add things that will stimulate visits during the winter. Shoot for consistent customer visits.

With all that background, here we go with a few ideas:

#1 PASTURED TURKEYS. This is a must for pastured broiler operations. Turkeys should not be a centerpiece because they are a seasonal item. But as a complement to a chicken operation they help extend the season on to Thanksgiving and Christmas while using the existing chicken infrastructure.

Do not underestimate the income generating potential of these complementary enterprises. If you do 500 turkeys at a net income of $15 per bird, that's $7,500 additional income without any additional investment in capital. When people come to the farm for the turkeys, they will pick up a few frozen chickens, rabbit, beef, root cellar vegetables or whatever.

Since it is near the holidays when people are looking for gift items, customers picking up turkeys are naturally more interested in buying potpourri, craft items and specialty baked items than they are any other time of the year. The festive mood is great for public relations.

In addition, people eat turkeys on festive, entertainment occasions, ensuring that in many cases your turkey will be eaten by someone in another family besides your customer family. This is a perfect way to acquaint more people with your farm.

#2 LAMB. Especially if you have cattle, sheep are a natural complementary enterprise. As a rule, you can add one sheep per cow without compromising cattle forage. In fact, it will actually thicken the grass sward, producing more forage than would have been possible with only one herbivore.

Lamb is a specialty product serving an ethnic market. Metropolitan areas with heavy populations of Middle Easterners are the high consumption localities. If you can access these ethnic groups you will be surprised at the potential market for lamb. For the next several years, all forecasters predict worldwide shortages of lamb but especially U.S. domestic production.

As with turkey, lamb consumption tends to be highest around special Jewish and Moslem holy days. Plan your production cycle accordingly.

The biggest drawback to sheep is the fencing requirement. An electric fence that will hold a sheep will hold a cow, but not vice versa. The fencing requirement and much smaller per capita consumption are reasons I did not include sheep as a centerpiece operation. But if you were to make sheep a centerpiece enterprise, you would certainly want to add some cattle as a complementary one since no additional fencing would be required.

#3 PORK. The market for alternatively produced pork is unbelievable. Per capita consumption is much lower than beef and chicken, to be sure, but it's hard to tell how much of that prejudice is being caused by highly publicized swine concentration camp abuses.

When we added pork several years ago we did not think our extremely health-conscious clientele would be interested. We added pork not because we thought we could sell it, but because it was a byproduct of our compost turning.

We certainly underestimated the market on pork. Customers we never would have dreamed would eat pork began buying whole hogs. The quality of pork raised on clean ground where these critters can root and eat forage is unbelievable. We would like to get to the point where the pigs self-harvest their feed from grain or vegetables (like zucchini or tubers) that we plant in the vacated ground they till by rooting.

Pigs are easy to control with electric fence but they are not as easy to herd as cattle or sheep. Extremely intelligent, pigs will make your life exciting. Outsmarting them can provide many moments of heart-stopping pleasure. Pigs absolutely do not have a bad odor if raised correctly and are one of the most interesting farm animals.

#4 RABBIT. You may not think too many people eat rabbit, but here again is a sleeper. Featured prominently in nutrition articles recently, rabbit offers the highest protein per pound without cholesterol. Rabbits store fat only in their viscera; that is why they are prone to heart attacks.

Preferred by many white tablecloth restaurants, rabbit enjoys quite a bit of freedom from government inspection because it is not considered an agricultural commodity. It is a high-priced item and capable of adding many thousands of dollars to your farm. Year-round production makes it a great cash flow stabilizer even though it takes very little land space. Rabbits are quiet animals and therefore especially neighbor-friendly.

Photo 11-1. *Showing off a litter of pastured rabbits, our 16-year-old son Daniel is producing and selling 1,000 rabbits this year. He began this enterprise when he was 8 years old and is now a premier grower in the mid-Atlantic region.*

#5 FIREWOOD. Certainly if you pursue the bandsaw mill idea and begin working with wood, a companion firewood business would be a natural addition. We sell a few thousand dollars' worth each winter during the low cash flow time. The season runs counter to our busy summer livestock season.

We sell about half our firewood delivered in a 2-ton dump truck which holds roughly five pickup loads. We deliver for free within 10 miles of the farm. It is unsplit but cut to usable length.

The other half we sell u-haul to folks who come out to the farm in their pickup trucks. This niche is strong and has the added advantage of getting some folks to the farm who may buy some eggs or meat during the winter. Being able to salvage the treetops from a logging operation or the odd pieces from a lumber mill makes this a great adjunct to those kinds of enterprises.

Certainly connected enterprises would be poles and posts from rot resistant trees.

Photo 11-2. *In many areas of the country, a sideline firewood business can bring in several thousand dollars a year while improving the quality of your forested acres. This is an excellent way to salvage otherwide low quality material.*

#6 AGRITOURISM/RECREATION. In case you haven't noticed, America is recreating. The entertainment business is huge, and growing. This is a result of more disposable income — or at least the perception of it — and subconscious dissatisfaction with life. The emotional consequence of hurried, harried lifestyles, fighting the expressway and sitting in a Dilbert cubicle is a yearning to taste something real, to experience something different.

Probably the best-known aspect of this farm business is the pumpkin patch. Almost every town has a couple of pumpkin patches where families can go and pick up pumpkins for Halloween. These are being marketed successfully to elementary schools and day care facilities for field trips. The most successful pumpkin patch enterprises now have a whole host of add-ons:

- petting zoo
- straw bale maze
- baked treats like cookies and cupcakes
- hay rides
- haunted house
- crafts and homemade toys
- miniature golf
- full food service

This can go as big as your imagination. Just to show you what can be done, I'll relate the story of a farm I know about. It all started one October when a gregarious, mischievous grandma dressed up like a witch during a Sunday afternoon family get-together and waved at passing cars on the rural road in front of the farmhouse. The next Sunday a steady stream of cars came down the normally quiet road, hoping to catch a glimpse of that "crazy woman."

The family suddenly realized how people craved entertainment. Although I don't know all the particulars as to what came first and how it all developed, over the next few years the family created an agritourism enterprise that supports several families. It looks like this:

- *Straw bale maze.* Using 3,000 straw bales stacked three high, they charge $1.50 a person to go through it. They employ local high schoolers to be up on top of the maze for security and to liven up the event. Literally thousands of people come through this each fall. Just to keep it interesting, these high schoolers will hop down periodically and shift an alley to a dead end by flipping some bales 90 degrees. Then when braggarts go through to show friends how fast they can get to the end of the maze, they are completely confused. But everyone loves it.

- *Observation deck* for grandma and grandpa to videotape their kids going through the maze. For 50 cents, folks can sit in this observation deck and see down into the maze to capture it on film or just to watch.

- *Miniature golf.* An old bank barn has been converted with traps and tubes to create a wonderful miniature golf experience. Of course, this costs $1.50 as well. Going down into the bowels of the bank barn is all part of the course.

- *Company parties.* During this time of year, many businesses are looking for a place to have a company bash for their employees. Renting this farm for an exclusive party for a 3-hour period now fills up the operations during the weekdays. Then the farm is open to the public in the evenings.

- *Catering.* These company parties always want a meal. Instead of hiring a caterer to bring the meal to the farm and serve it, these enterprising farmers now cook the meal for additional income. This is a way to move their beef cattle and eggs at a high value, expose hundreds of people to their meats and vegetables, and simplify the logistics and planning for the bash by the company officials.

- *Pumpkin patch.* Instead of having people walk out into the pumpkin field, these folks mow a nice 3-acre paddock near the road and illuminate it with red filtered floodlight. This accentuates the oranges and green grass, creating a special effect. Children can walk on the grass without tripping over vines. I think now these folks do not even grow their own pumpkins, but buy them by the tractor-trailer load. Sized small, medium and large, the pumpkins sell for different prices.

- *Face painting.* For another couple dollars, children can have their faces painted by local high school students employed for the purpose.

- *Concessions.* Easy items like candy apples and snow cones create additional income-generating opportunities, as well as contributing to the festive atmosphere. These folks try to stay with healthy type treats to maintain a family-friendly, health-conscious perspective. Carrot sticks and healthy chips add to this aura.

- *Arts and crafts.* Local crafts are available for sale at the big barn and this provides wonderful supplemental income for neighbors. Sold on commission, this brings in even more money to the farmers.

- *Hayrides.* For an additional fee, you can enjoy a tour through some woods. I know of some farms in this business that spice up the hayride with scary elements. Plastic streamers hanging from tree branches can make the experience more exotic.

- *Haunted house.* Some outfits like this have been extremely creative at doing a whole host of things to stimulate fear. I would not encourage anyone to violate their conscience over this whole tactic, but for some it has worked well. For example, sending people across a black plastic "Peanut butter swamp" and then encountering a person at the end who jumps up out of a pit with a chainless chainsaw raised and revved to chase everyone to the next scare item ensures squeals and memories.

- *Bonfire with marshmallows.* For 50 cents more, kids can skewer a marshmallow on a stick and have a "s'more," that tasty delight created by sandwiching a freshly roasted marshmallow between a section of chocolate bar and two pieces of graham cracker.

As you let your mind run with this, the sky is the limit on additional moneymaking facets. Although these folks live in an extremely rural area, they get tens of thousands of people to their farm every October and it is a full-time income.

I would include in this section about agritourism things like farm and ranch vacations. An entire network, similar to the bed and breakfast directories, exists to offer alternative farm-type recreational opportunities for people. Many of these are dude ranches and offer three full meals a day as part of the package. But others are quite simple with no meals offered. Certainly if you have an interesting farm you have something city people would enjoy seeing.

Although I know several farms for which agrirecreation is now the centerpiece, I know of no farm for which this originated as the centerpiece. All these farms that have successfully developed agritourism as a full-time enterprise started as dairies, livestock farms or other more conventional enterprises and saw this as an opportunity to generate additional income. I think this can become the centerpiece, but usually a farm does not start with this in mind.

#7 GREENHOUSE/FLOWERS. Especially if your centerpiece is a market garden, a greenhouse producing bedding plants or extending the season is a perfect mate. In addition, flowers both to transplant or as cut flowers can mesh nicely with a farmers' market booth or your own on-farm sales enterprise.

By not trying to make this a centerpiece enterprise, you can get along with a poor-boy greenhouse, which completely changes the capitalization costs and the potential payback period. By now you're beginning to see why some things that make excellent complementary enterprises are not as sensible as centerpieces.

Depending on your local market and production expertise, sometimes an enterprise like cut flowers can overtake the market

Photo 11-3. *Betsy and Amanda Womack enjoy working in their family's greenhouse, especially when early spring days are blustery outside. Bedding plants and early vegetables offer exceptional supplemental income.*

garden. These changes certainly do happen. I know farmers who thought they were going to be full-time herb producers and began growing a few perennial flowers on the side. In a couple of years, the herbs were the little things on the side and the flowers had become the centerpiece.

Staying flexible, watching the market and listening to your own intuition often make your enterprise change over time. The income mix you envision today will probably not be the reality five years from now. Thirty years ago pastured broilers never figured into my imagination. Who would have guessed it would become our centerpiece?

I enjoy asking college graduates if they are working in the field of their undergraduate major. Very few are. This is because we grow and change. The same is true for farm enterprises. Where you start may well not be where you end up. The important thing is to be perceptive enough, and loose enough (including carrying a low debt load) that you can make the changes necessary to capitalize on the current next best thing for you.

#8 HONEY AND BEE PRODUCTS. This includes everything from pure honey to beeswax candles or figurines. I have a friend who made a bunch of beeswax nativity scenes for Christmas.

Good local honey is becoming more difficult to get. Mites play havoc with hives and plenty of apiarists are scrambling to keep hives healthy and productive. Honey is a clean and green, natural product that is being carried along with the whole organic foods movement.

While a full-time beekeeping enterprise is a major undertaking, picking up a few extra thousand dollars with a small enterprise is quite doable. This is an excellent adjunct to any centerpiece enterprise.

#9 LAND MANAGEMENT — STOCKERS. In places all over the country abandoned land is ready to be utilized. And I do not mean exploited. I mean cared for, stewarded, and loved. With the aging farmers comes land without warm bodies, without care. And with absentee ownership comes ignorance about how to properly care for land.

The upshot is that a huge amount of land is available, especially open land. It is reverting to brush and forest when it should be lush perennial prairie polycultures. The Native Americans and lightning maintained this balance with fire. I recently heard about an Arkansas University graduate student who drove across the state of Alabama looking for unused farmland. When he spotted acreage that looked abandoned, he would go to the farmhouse and ask if it could be rented. He stopped at 100 farms during his multi-day search. At each place, his pitch was the same: "Hi, I'm John Doe and I wonder if you would consider renting your land to me?" Not one time — not once — was he rejected. Every single owner — in most cases an elderly couple or single — welcomed him and said they would be glad to rent the land. What he found was that the owners seemed genuinely grateful that someone had actually noticed and taken an interest. These were cold calls from a total stranger.

For land to be productive, we must maintain it with stewardship, concentrating on livestock as a land management tool. Utiliz-

Photo 11-4. *Using livestock as a landscape management tool, returning perennial prairie polycultures and reducing fossil fuel usage on grasslands — these are all possible with intensively managed stocker cattle. This is rented acreage a few miles away. A realtor says the land and vegetation improvement warrants a landowner paying the farmer.*

ing the principles outlined in my book *Salad Bar Beef,* you can take advantages of this available acreage and turn handsome profits.

Since farmland rental is based on the cultural notion that "there ain't no money in farmin'" it is possible to access large acreages for very little money. Intensively managed stocker cattle, while not returning a large income per acre, can give a handsome return per hour.

Rather than a direct market enterprise, this is a commodity program wherein you buy and sell wholesale. That makes marketing easy. If you can get access to land, stockers can net $5 per acre per inch of rainfall. The lower the rainfall, the lower the per acre return. The higher the rainfall, the greater the per-acre return because the greater the potential forage productivity.

Realtors have told me that they have new farm landowners asking them for the names of farmers who can take care of their acreage. These folks come from the city and buy their rural retreat but they want to maintain the viewscape, the open vistas. That takes either a couple of mowings annually, which is expensive and does not put the land into a lower tax bracket, or it requires grazing livestock.

The way 95 percent of livestock are grazed in this country the average new farm owner understandably does not want animals on the place. But with intensively managed grazing, we can actually beautify the property, reduce weeds, fence out riparian areas and enhance the wildlife populations.

The basic idea here is to buy roughly 500 pound calves, keep them 6-7 months and then sell them in the fall. We use this model and it fluctuates from year to year, but on average we can figure the above net return after all expenses, including rent, are covered. For the other months of the year, we have no cattle on the property.

It requires no machinery and no buildings. What it does require is some water pipe, electric fence, cattle, grass and experience. Therein lies the problem. I guarantee that I could go onto any farm in our county and double the production in one year just with grazing management. You could too, if you had the know-how.

The weak link in the entire farming business is the lack of people who know how to make a living on a piece of land. The land is there. The cattle are there. The hardware is there. But we're in dire need of people with the experience to put a profitable enterprise together. All most farmers know how to do is pay bills, complain and do what the neighbors do or follow the advice of the extension service.

After doing a couple of these land management projects with stocker cattle, I am convinced that in some cases you could actually get paid to put animals on the property. You could peddle a management portfolio just like any salesman peddles his aluminum siding job, with pictures or whatever. You get clients and run the company as a nostalgic enterprise: return the dynamics of buffalo and prairie (cattle and grass). The opportunities are enormous.

#10 COTTAGE INDUSTRY. This is a catchall, and I can't begin to list all the possibilities. Every time you think you've heard of everything that could be made and marketed, you hear of another dozen.

I ran into a fellow in Illinois whose daughter wanted to have a full-time home-based business. She decided there was a real market for little baby decorative straw bales measuring about 2" x 4". Used in craft displays, these little jiggers are a hot seller. Her dad found blueprints for a demonstration baler originally used by one of the farm machinery companies when balers were first coming out. A local machine shop built one. Her grandfather cut the little wires to tie the bales and she began making bales. Do you know that within two years she was grossing nearly $30,000? Who would have guessed? What amazes me is that in every nook and cranny of this great land, creative success stories like this exist. Yes, some, as in this case, do become big enough to be centerpiece enterprises. But even this one was a spin-off from the family farm.

A farm or homestead gives you a place to launch a cottage industry. This was the way the world functioned for centuries up until the industrial revolution. Shoemakers and candle makers had their garden, milk cow and pig out behind the house in the village. Their crafts functioned in the economy partly because the household's cash expenses could be kept low with the food produced there.

Often a cottage industry can become a centerpiece, but one of the best ways for a cottage industry to thrive is to keep it unsaddled from high domestic living expenses. If the fledgling enterprise must pay for two cars, a mortgage, all food costs and utilities, it generally cannot generate the cash flow to stay viable. But if, through your viable farm, the cottage industry does not need to pay your living expenses, profits can be plowed back into it quickly to make it downright lucrative.

Being able to plow the cash back into the enterprise and not siphon off before-expense dollars to pay non-business bills frees the cottage enterprise to grow at maximum speed. If your enterprise really takes off, the more attention you can spend on that business, the better. The safety net, however, is the vegetable garden, the chickens scratching in the backyard, a full woodshed and the herb garden outside the back door. This provides the incubator and catalyst to launch the cottage industry.

Perhaps one of the best sources for stimulating your creative juices in this regard is Ken Scharabok's *How To Earn Extra Money in the Country.* Just to give you a feel for this book, here are just the Ps from the table of contents:

Packaged Rabbit Manure
Painting
Pantyhose Dolls
Parking Lot Maintenance
Part-Time Attorney
Parts Distribution Point
Part-Time Real Estate
Part-Time Truck Driving
Party Maid Service
Pastured Poultry for the Organic Market
Pawn Shop
Peafowl
Personal Computer Consultant
Personalized Trading Cards
Pet Accessories, Treats or Toys
Pet Cemetery
Petting Zoo or Farm
Photograph Album Consultant
Pickup Truck Caps
Pine Cone Harvesting
Preparing Income Taxes
Primitive Campground
Private Airport
Process Server
Producing Sweet Sorghum Syrup
Professional Deer Hunter
Professional Fund Raiser or Event Coordinator
Property Caretaking
Portable Bait and Tackle Shop
Pumpkin Farm

Remember, these are just the Ps. Some ideas may seem far out, but that's okay. If you need your creative juices stimulated, Ken's book will do it.

The whole idea of this chapter is to present viable add-on enterprises that, taken as a whole, will provide a full-time living. To make a full-time living from any one is nearly impossible, but a couple thousand dollars on this one, and a few more on this one and a few more on another one eventually add up to a full-time living.

Often, these complementary enterprises can be the center-pieces for your children, for tomorrow's generation. Once you realize how many enterprises can be stacked on a land base, you realize how foolish we farmers are to think that we need more land in order to employ our children. We don't need land; we need creativity. Multi-generational, high-income agricultural/cottage industry complexes truly offer a farm life that is environmentally, emotionally and economically enhancing.

Ten Commandments for Succeeding on the Farm

#1 STAY AT HOME. You're not living in town anymore. If you want town life with little league and ballet, social amenities and movies, then stay in town. But if you're going to live in the country, and especially if you intend to farm full-time, park the car. Act like a person who lives in the country.

Your farm is your life and your work. Trips to town are costly, not only in time and money, but also emotionally. I'm more tired after half a day fighting traffic and cash registers in town than I am after making hay all day.

If you don't like to stay at home, then probably you should just shut down the farm, but don't say that farming doesn't pay. This is one of the most irritating things to me, to hear people blame farming for not being profitable when all you see is plumes of dust rising up behind their car because it's ripping down the road again.

The farm is your recreation, your life, and your work. It should energize you and be your place of contentment. Few things are more incongruous than a three-car farm family. I never could figure out why farmers have to go to theme parks or go bungee jumping. I get all the excitement I need watching trees fall or sorting cattle. Watching a perfect patch of strawberries ripen isn't far behind, either. Certainly picking them off the vine and eating them "on location" is close to ecstasy.

I hope I'm not misunderstood in this discussion. Don't think I'm suggesting every farmer needs to be a hermit "stick-in-the-mud" sequestered from society. I certainly enjoy art, drama and a good concert. But a huge difference exists between partaking of these enjoyments as a special occasion and being the city socialite. Plenty of stimulation and opportunity exists in the "special occasion" spectrum to whet the appetite of our children and spouses.

I've watched too many farm families flounder on this issue. A husband or wife wants to be involved with every committee or shindig that is listed in the newspaper's "What's Happening in Town This Week." Perhaps the children want to attend every school athletic event, or every rolling museum exhibit that comes through town. We only have 24 hours in a day, and if we fail to recognize the cost of being away from the farm, we will fail.

The successful life is not one of frivolity and haphazardly flitting from here to there like fairies in tall tales. Rather, success will take discipline in all areas of life. Remember, too, that we are talking about what it takes to start. After the farm is established and turning a good profit, then you will be freer to be away. Our family goes away for weeks now each year — perhaps making up for lost time. But if we had started doing that, we never would have been successful.

Everything has a season. Life, business, family: all of these have seasonal characteristics. In the early days, you will find the old saw "keep your nose to the grindstone" better advice than just about anything. Later, you can enjoy the fruits of your labor.

I'm not suggesting that you can never go away, or that recreation is completely out of the picture. But you can do many things — like picnics, horseshoes, swimming or impromptu football playing — that do not pull you away from the farm.

#2 BE HONEST ABOUT LUXURIES. Separate business and pleasure. I always cringe when I hear people say they want to farm and the first thing they want is a horse. Mark it down, they're on their way out. Recreational horses have to be the single biggest profit sucker in farming. If the farm is making money, that's a different story. And when I say making money, I mean providing a decent full-time income.

Second cars, diesel pickup trucks, fancy vacations and designer labels can all come AFTER the farm is profitable. Every unnecessary item you buy, especially in the early days, will delay or destroy your chances of ever being successful at farming. To dismiss these luxuries as just "little things" or inconsequential may be humorous among your friends and relatives, but it's no laughing matter when you're frustrated over the farm not giving you a full-time salary. Other items that often take more money than you think are eating out and trying to keep the house a cool 70 degrees in summer and a warm 80 degrees in winter.

Again, I'm not suggesting we should live with the same privations as the Jamestown colonists experienced, but neither do we need to live like kings. A rusty pickup in sound mechanical condition will carry just as much as a dually tonner. One costs $2,000 and the other $40,000. "Keeping up Appearances" is not just a humorous British sit-com; it's also very expensive.

#3 BUILD RELATIONSHIPS WITH NEIGHBORS. Don't come into the community like some arrogant know-it-all and start flailing away at the neighbor who has a feedlot or a manure lagoon or who sprays his corn with pesticide and buys truckloads of fertilizer.

Don't complain to the neighbor about his two junk cars propped up on blocks or the unsightly trash pile right up against your fence. Let it never be said that you spoke a demeaning word against a neighbor.

Rather, be friendly and invite these folks over for get-acquainted times. Help them. Perhaps you could help the big farmer spread slurry. Even though it's reprehensible philosophically, you will never reach him by being vituperative and unfriendly.

Don't spurn your neighbors. I've watched newbies move into conservative farming communities and whine: "These folks aren't very friendly. They haven't invited us to dinner yet. They don't include us in their social events."

My question is this: "Have you? Are you going to blame them for your behavior? Get real. He who wants friends must show himself friendly."

Don't complain about acceptance in the neighborhood if you haven't gone out of your way to be friendly. And that means, of course, keeping your own nuisances down. Keep your animals at home, and that includes the dog and the cat as well as livestock. Get over on the narrow road to pass. Slow down on dusty country roads.

Be patient with your endeavors, because these relationships take years to build. And even if your neighbors never become your best friends — if you're different, they probably will not — you want to at least have a good enough relationship that you can call on them in crisis and get help. If they call on you for help, then you know you have arrived.

4. INVEST IN MONEYMAKING THINGS. A livable house is all you need. A car that will get you there is all you need. You don't need to live in style or drive in style. That can come later.

Stay focused on specific items that will stimulate cash flow to make the farm more economically viable. Remember function over form.

I just can't stand it when people say they can't make a living on the farm and their trashcan is full of soda cans, frozen pizza boxes and empty bags of dog food. Do you realize it's easy to spend hundreds of dollars a year on dog food for a big dog, not to mention vet bills?

Yes, I know the poor critter is just lovey-dovey and looks at you with big sad eyes. Around here, all pets must carry their weight. Our cats do not get cat food. They are our rodent control and when they get old and can't live off the land, then they die. We don't spend money on them at the vet's. That is not cruel; it's keeping things in the right perspective. Our cats are as healthy as any I've ever seen, but they are still cats.

A guard dog can certainly pay his way, but only if you have a predator problem. Let the pets be of necessity, not just because they're fun. Rachel, our daughter, enjoys horses. We're glad for her to have a horse as long as it pays for itself. That may sound harsh, but the alternative is a few hundred dollars here, a few hundred dollars there, and suddenly the farm is not paying the bills.

#5 BE COMMITTED. Paul Harvey says that what separates successful people from failures is that the successful ones get up when they fall down. Be assured, you will fall down. The question is, will you get up?

As long as you get up one more time than you fall down, you are successful. When you have a setback, learn from it and go on. Be persistent if you think something has possibilities. Stay with it, to work through the learning curves, until it becomes profitable. Realize that all the time you're going through the process, you're becoming more proficient and this will make your increasing with the market easier.

Just how badly do you want to succeed? Can you taste a full-time farming enterprise? Can you feel it? Can you visualize it? Can you smell it? Can you hear it? This will help you stop making excuses about this stupid animal or these dumb plants. They are acting exactly the way God made them. It's up to you to modify your production model to capitalize on their instincts, to get them under control, and make them profitable.

#6 DO SOMETHING. Even if it's wrong, do something. It's in the doing that you discover the right approach. You'll learn more from your failures than your successes so get busy and fail a few times. With each failure you'll zero in a little closer on the right option. Don't just scratch your head and sigh.

And don't complain about having bad luck. Everyone has bad days. Realize the key to your success lies between your ears and get with it. As you move, you'll see over those little hills in the road and know better which path to take. The knowing is in the doing.

#7 WRITE DOWN A PLAN. A corollary is the holistic management approach: have a single goal. If you cannot boil your mission down into a single goal, you will be plagued with schizophrenic thinking the rest of your life. And a push-me-pull-you type of farm will never be profitable.

Something about going through the physical exercise of writing down a plan, writing down objectives, writing down action steps, helps you to focus in a way that nothing else will. Keep that plan posted on the refrigerator or the desk, above the bathroom sink or at the headboard of the bed. It is your ticket to dream fulfillment.

#8 FULLY UTILIZE YOUR RESOURCES. Certainly creatively capturing your resources is easier as your exposure level to ideas broadens and as your own success increases your confidence. But be careful about categorically dismissing things as being unusable. Many a junk pile is the beginning of invention.

I've gone onto farms and seen dilapidated buildings the owner labeled "eyesores" that could be turned into a farm and ranch vacation cottage, a craft sales building or drying shed for flowers and herbs. Yes, times do come when buildings may be unsalvageable (almost any bank barn qualifies) but if it's not in the way, let it sit. You may hit upon a real use a couple years down the road and then you'll kick yourself for not saving that building, pile of boards, or mass of half-inch pipe.

Think about gravity and water, south slopes and solar, earth and 50 degrees, neighbors and rotting round bales or nearby cities and chipped Christmas trees. Inventory what are perceived to be problems for others and see if you can use these things.

If you're next to a hunting preserve that buys 200 pheasants a week for $10 apiece from someone 200 miles away, consider capturing that resource. Too often folks move into a community and don't realize the material and human resources right under their noses. Go looking and thinking about how you can capitalize on all these resources.

#9 SURROUND YOURSELF WITH COMPETENT COUNSEL. Spend time with people doing successfully what you want to do. Don't waste your time with naysayers and grumblers.

Not getting bad counsel is just as important as getting good counsel. The one is as damaging as the other is enlightening.

Instead of going on vacation, go visit a successful practitioner. Invite a customer family to dinner one night. You may be surprised to find a competent accountant, attorney or marketing specialist. Join the sustainable agriculture organization(s) in your state and develop friendships there. If you want to grow grapes, go spend a few days with a vineyard owner, pruning or toting, just to learn what you can. If you want to produce beef, go help a good cattleman for a couple of days. In my view, you cannot overexpose yourself to good advice.

#10 BE CONSISTENT. Don't start and stop endeavors, but offer the product over the long haul. Establishing a clientele requires consistent delivery of service or product. On again, off again enterprises never get anywhere.

If you're going to produce eggs, then devote your attention to never turning someone away. Okay, so a predator got in. Perhaps the birds got cold and stopped laying a couple of days. Yes, that is all part of the learning curve. But try hard to overcome these glitches so they don't happen again.

Don't be maybe and maybe not. People will not buy things unpromised. If you say you will have grapes for September, people will order. But if you say you're not sure just what you're going to do with grapes, people won't order.

Do whatever it takes to fulfill your promises. I've slept in the chicken house, waded through floodwaters at 2 a.m. with a bale of hay on my back while Teresa held the flashlight. You do whatever it takes to make good on your product, and the market will respond accordingly.

If you practice these principles, your farm will gradually show a profit. The closer you stay to these ideas, the sooner your enterprise will be weaned from outside capital and the sooner you can begin to violate some of these without jeopardizing the entire operation. It's fun to get to a point in your profitability that allows you to savor some of the things you deny yourself early on. The pinnacle of success is worth the early discipline.

Embarking on your Venture

Acquiring Land

Although farming does not require land ownership, having a place you can call your own, that you can invest your life in and then leave better than you found it is certainly a noble goal. Speaking of the perfect world, the *Bible* describes each man "sitting under his own fig tree." Although we should not construe this beyond its meaning, it certainly enjoins land ownership and private property rights of some magnitude.

As a beginning farmer, aspiring to own land is not wrong. It is noble and good. And although I've spent ample energy trying to make sure that you get this goal in the right perspective, it is nonetheless the ultimate of the farming experience.

Here are some prerequisites to purchasing land. Ask yourself if any of these have been met.

★ Have a viable patron base already purchasing what you are growing on land to which you already have access. This can be your own backyard, borrowed or rented land. The point is that you have a clientele that will follow you to your new location and provide a loyal marketing base to get a good profit on the very first thing that comes off your acreage.

★ Know how to make money on land. This, of course, presumes you've been doing something long enough that you feel comfortable risking more than just sweat. When you begin risking your pocketbook, things get serious real fast. Your experiential confidence level must be high. This does not include ideas you've gotten out of books or cooperative schemes you've picked up from magazine ads. If you've turned good money on a piece of land you've been working with, chances are you can duplicate this effort in similarly rewarding fashion on one you own.

★ Have enough money to live on for 5-10 years *after* you've completely paid for the farm. I don't mean contracted for, I mean paid for. This one gets sticky. If your experience and knowledge level is low, your learning curve will be steep. You'll make more mistakes than you imagine possible. Don't worry, that's normal for any fledgling business.

Do not miss the admonition "after" you've paid for the farm. Here is a typical scenario:

Man is financially successful.
He lays money by.
Begins dreaming of being gentleman farmer.
Begins looking at real estate ads and puts "feelers" out on grapevine.
Looks at some land.
Has enough equity to make sizeable down-payment.
Sells house, puts equity in farm, and borrows rest.
Needs job to pay debt.
Too tired from off-farm job to do anything with farm.
Weeds grow -- generally unproductive.
Honeymoon over, begins feeling frustrated.
Trapped with big debt, thinks farming doesn't pay.
Dream vanishes.
Puts farm on the market.

The average person who buys a farm today will have that farm back on the market within five years. This cycle is duplicated over and over and over again, and I don't want you to go through it. Life is too short to be complicated with this.

I've seen this cycle repeated too many times. If you have enough money to buy the land debt-free, you won't need to generate the cash to pay for it from your outside work. I cannot overstate the importance of this issue.

You can buy a couple of acres, move in a $3,000 used mobile home, and get started for perhaps less than $10,000. If you are debt free, you can work part-time — even substitute teaching — while you develop your farm.

But if you have a mortgage to pay, even if it's only $3,000 per year, it will saddle you with cash requirements that will push you into a more demanding off-farm job, which will leave you less flexibility to develop the farm. The magazine *Countryside* has featured testimonials time and again from folks who maintained a family on less than $500 per month. My own experience was like this.

The best money you can make is what you save on living expenses. Get rid of the second car, quit patronizing the grocery store, and do more for yourself. If you grow most of your own food, harvest your own heat off the farm (firewood) and get clothes at the thrift store, you can live quite cheaply. As the farm becomes lucrative, you can begin to upgrade. You will see what you can more economically buy off the farm than grow yourself, and you can begin focusing more energy on the things that generate a high return per hour.

But you must keep your cash requirements low. If you're going to move onto the acreage, build a nice, acceptable house with well-mowed lawn, you'd better have deep pockets. In such a case, if you can't live farm-income free from 5-10 years, forget it. Why so long? It will take you that long to go through the learning curve.

Business guru Peter Drucker charts the business cycle and points out that any new business has a 7-10 year period before it becomes lucrative. There just aren't any shortcuts to this no matter how you may think folks that are there got there easy. None of us did.

And you won't either. In his wonderful book, *Guerilla Marketing Attack,* Jay Conrad Levinson stresses that the single most important personality trait of the successful guerilla is *patience.* You see, there's nothing wrong with having a nice house and two cars, but timing is everything. If you try to rush it, you may never attain what you want. The old adage: "Good things come to those who wait" could not be truer. The farm income must push the upgrading, or the upgrading will ruin the farm.

It's really that simple. I've upgraded to the point where I don't even change my car oil anymore. When we started, and didn't have cash, I wouldn't have thought about paying someone else to do it. But now that my farm time is so much more valuable than shop time (because I'm not a good mechanic) it's ridiculous for me to spend two hours doing something I do not enjoy when I can pay $20 in town and get it done. And the folks at the oil-changing business will grease, check the differential and transmission, all the engine fluids. It would take me forever to jack and grunt and get all that done.

But that is what I call "upgrading." You don't start there. You start down in the trenches, saving every penny, and staying focused on keeping your income running ahead of your expenses. As the income runs farther ahead, you can begin spending a little more. But if you start strapped with high expenses, where you must generate cash flow to keep the operation solvent, it will never go. You simply can't turn the cash fast enough early enough to make it go. That is why I am concerned that most folks who want to farm will run out and buy land, and be in that vicious cycle until they burn out.

Well, suppose you've passed enough of these prerequisites to justify buying land. What then? First, decide on a general area where you'd like to be.

Believe it or not, a healthy percentage of farmland changes hands without going through realtors. It's done through connections in the community. Begin developing contacts in the farming community. Visit the livestock auction barn regularly. Attend farm meetings in the community. Get friendly with the extension agents – this is where they can be helpful, because they certainly get around. And they get around to a lot of farmers who are on the verge of collapse because they have been listening too much to free advice.

Join the Farm Bureau Federation. Basically, get involved with the conventional agriculture community, because that is where the land is. And that is where the land becomes available. I've watched patient folks get great deals. Sometimes it will take a couple of years, but again, patience is the most important virtue. Be willing to wait until the right place comes up at the right price.

Buy something within 40 miles of a 25,000-plus population center. That is not a big city, but it would knock out large areas of the West. It is generally better to spend more per acre for land closer to town. Because marketing becomes easier the closer you are to town, put your land money into location rather than size.

Look at it this way. If you have $50,000 to spend on land, it would be better to buy 10 acres on the outskirts of a metropolitan area than 100 acres in the middle of a desert. You can generate $100,000 on the 10 acres when you probably can't generate $2,000 on the 100 acres.

Always check for water. Water is the most basic element of life; without a good source of water, you will never be able to reach your potential. I remember well telling an elderly neighbor one day about a friend who was getting ready to buy a small farm. I was describing different features about the land and the old farmer interrupted me: "Does it have good water?" When I told him about the pond on the place, he went on: "If you ain't got water, you ain't got nothin'." He's right, of course.

If you outgrow your small farm, if you practice the principles in this book you will be making enough money to buy a bigger place not far away. The more we do here on our farm and the more creative folks I meet, the more impressed I am that the amount of acreage is a small matter. If you're willing to intensify your efforts, and to diversify, you can earn big bucks on small acreages.

I encourage folks to buy poor land rather than real good land. You can capitalize on the mistakes of others and turn poor land around.

But if you sink everything into real good land, you may not have anything left to make it generate cash.

Look for diverse ecosystems within the acreage: a pond site, some woods, some nice level ground. But realize that if it doesn't have those, you can add them at a later time. That is the principle of stewardship — you can make it more diverse and far more productive than it is in its wild state.

By the time you are ready to buy land, you should have a good idea of what you want. Write down things you're looking for and just match parcels to that list. If you don't know what you're looking for, you're probably not ready to buy. The world is full of folks who have more money than sense.

Remember that the way to big acreages is through small acreages. You can trade up and invest up out of the profits generated on the small acreage. If you're frustrated on a small acreage, you'll be more frustrated on a large acreage. Refine and fully develop what you have before looking for something else.

And then repeat the same procedure there. By the time you get a larger acreage, you'll have the experience and the know-how to be a good steward. If you get too large an acreage too soon, you'll be a poor steward and that doesn't do anybody any good.

Land should only be acquired when you know what to do with it, and the size should be less important than location. Be patient and let your farming enterprise drive the land base, rather than the land base driving the farm.

Where to Settle

"You really live in the sticks!" We're used to hearing the exclamation as people from around the country come to visit the farm and expect that because of our customer base and marketing success we live on a major thoroughfare. Nothing could be further from reality. We live in the proverbial boondocks on a dirt road. But, and this is a big "but," we live only 15 minutes from Staunton, a town of 25,000 people. Our county, with its two towns, has a total population base of about 85,000. This is not a metropolitan base by any standards, and yet it is enough. I agree with Ralph Waldo Emerson's famous observation:

> "I trust a great deal to common fame, as indeed we all must. If a man has good corn, or wood, or boards, or pigs to sell, or can make better chairs or knives, crucibles or church organs, than anybody else, you will find a broad, hard-beaten road to his house, though it be in the woods."

No paradise exists this side of eternity. Every place has assets and liabilities. The more remote your location, the lower the cost of living but the more difficult it is to market your products. The more urban your setting, the higher the cost of living, including land prices, but the easier it is to find customers.

Certainly you should be somewhere that is compatible with you. If you break out in hives every time the temperature hits 90 degrees, Florida is probably not the place for you. If you break into chills at 75 degrees, you may want to look real hard before locating in Bismarck, North Dakota. Find a place you like, but don't be married to a place because no matter where you go, you'll find things you like and things you don't like.

Remember, you may call yourself a farmer, but you're really in the people business. And from place to place, people are more similar than dissimilar.

[Let me offer a disclaimer at this point: all my advice is generic and exceptions do exist.]

The cheaper the land, the more rural and the harder to market. In addition, the more expensive it is to purchase supplies. A trip to town to buy hardware, fencing or a tractor part is a major investment when it entails a one or two hour drive. That can quickly offset the low land prices.

In addition, the indigenous population in extremely rural areas tends to be less interested in alternatives. They are more traditional. Not that tradition is a bad thing, but it can inhibit exploring positive changes, like locally grown organic produce instead of cheap Mexican imports. Or the notion that better food might cost a little more. Or the notion that there is better food.

A cosmopolitan awareness opens up more opportunities for the alternative farmer. You need neighbors who realize that supermarket chicken is 10 percent fecal soup; who know that conventional sweet corn receives a pesticide application once a day to keep it bug free. You need neighbors who realize there is more to life than chewing, spitting, drinking and watching TV, even though nearly every community has a healthy contingent of folks who partake of these pastimes.

Settle where you like the landscape and the people, where you have connections, and where you think you'll receive support for what you want to do. By all means do not think that the market niche for specialty food is close to being saturated.

A couple of years ago a friend who lived only 12 miles away was interested in duplicating our pastured poultry model. He was concerned that our rural community would not be able to support two of us with the same product. Neither did he want to take customers away from us and risk bad feelings. He came to me and asked what he should do.

I told him to spend the next week asking people at the bank, at the store checkout line, on the street — anybody he happened to meet — if they had ever heard of us. You must understand, we are fairly high profile in our area. I have no idea how many times we've been in the newspaper, on radio and television because we're the resident commercial alternative farmers. We've even been featured in *National Geographic!*

In addition, I routinely write letters to the editor on agricultural issues. Every time a major food contamination story hits, here comes the press, taking pictures of clean animals. It all amounts to quite a pile of free publicity. Anyway, he spent the week doing exactly what I suggested.

At the end of the week, he came by and I asked him what he found. Shaking his head in complete disbelief, he said he did not find one person who had ever heard of us. Isn't that amazing? The opportunities out there for clean food are beyond description. The only thing holding back the market is the lack of farmers who will make truly distinctively better food more available. As soon as people try it, they're hooked. More producers will simply snowball the awareness and increase the social consciousness until finally people will demand better food.

I *have* seen people inhibited by location, and almost always it is because they are in too remote an area. They settled there to "get away from it all" and because land was cheap, but they will spend the rest of their lives trying to make marketing work. I've said this before but it bears repeating: I'd rather spend $100,000 on 25 acres thirty minutes from a city than the same amount on 1,000 acres in the middle of a desert. The income-generating potential will make the more expensive land by far the cheapest.

Furthermore, if we just forget about land ownership, and think about opportunities, where people are will be where the biggest needs are. And the more needs, the more niches for you to fill. Landed elderly folks needing groundskeepers are far more numerous around populated areas than they are in Montana. Unused acreage surrounds every city.

Close to cities is close to transience, and transience means opportunity. People are coming and going, leaving opportunities in their wake. Compare that with owner turnover in large ranch areas and you will understand my point.

If your farming business consists primarily of a customer base and product, rather than land, it is mobile and can go anywhere. I know people who have developed highly successful pastured poultry enterprises on unowned land, only to have their land base change. They just load the pens on a trailer and head down the road to the next location, kind of like beekeepers move their hives around from place to place.

A customer base is highly portable within a 50-mile radius. Whether you grow cows on the north side of town or the south side is immaterial. People can drive one direction as well as another for their food. This is one of the beautiful things about direct marketing. But as soon as you begin building silos, manure lagoons, concentration camp confinement houses and installing underground water systems, you will do anything to keep from changing locations — even go bankrupt. Building mobility into our farming enterprises is highly desirable, extending flexibility beyond our generation to our children's generation.

Allan Nation, along with Tom Lassater of Beefmaster fame, prophesy that we are entering an era of landowners and land managers. They see a division occurring as wealthy people invest in land as a way to maintain wealth. It is almost impossible to generate wealth by purchasing land because the return on investment is too low to maintain a salary in addition to the return. But land ownership is an excellent way to maintain wealth — a defensive posture.

Acquiring wealth, however, is an offensive posture and is generally best done by renting. Let's illustrate the point with a 100-acre grazing farm. In this area, that farm would sell for $250,000 and rent for $2,500. If we can put on 100 stockers yielding a gross margin of $150 per head (the difference between purchase price and sale price) the total gross returns are $15,000. How long would that take to pay for the land? The mortgage alone, at 8 percent interest, would be more than $20,000 a year. But if we rent it, we have $12,500 left over to pay salary and expenses. Renting can take many forms and you will find a deal that works for you, but the important thing is to understand that renting is an offensive wealth accumulation posture and in order to be fully utilized requires a mobile farming operation.

Renting seldom lasts a lifetime. Obviously you want to negotiate the best terms possible. The big variables are:

- Length of time guaranteed in the lease
- Who takes responsibility for fencing
- Who takes responsibility for fertilizer, weed control, and general landscape issues
- Minimum requirements, from a landscape aesthetic point of view, on the renter
- Renter's freedom: cutting down a tree, changing an internal fence, etc.

Some of the greatest opportunity exists in row crop areas, and especially in cotton country. Anyplace where farming is depressed due to the industrial model is a good place for you to come in and pick up the pieces. Last year I was in Iowa where this county's population had dropped steadily for the last 10 years. Land that sold for $3,000 an acre in 1978 sells for $1,000 today — and rents for a lot less. The Mississippi delta and all the post-cotton areas of western Tennessee and Kentucky have inviting land opportunities.

If you have the right farming mindset, you will see opportunities virtually everywhere.

A man in New England built portable hoophouse henhouses and tows them down the road at 30 miles an hour from place to place to follow his cattle. He can have a rental piece here and one over there, but everything can move from place to place. We need to focus on maintaining flexibility, and then as the profit builds we can finally think more about a permanent place. But initially, the more portability we can build into the enterprise, the more options we can keep open.

All this boils down to time. The more time you have, the better your decision will be. You simply cannot make an honest assessment of a place by visiting a couple of times and then plunking down a downpayment on land. It's much better to go and rent and take a job somewhere to buy yourself time to explore and learn.

The common fortune in Chinese cookies is true: "Good things come to those who wait." Impatience is perhaps one of the single biggest destroyers of would-be farmers. This puts an emphasis on this settling down issue that is quite different than most how-to advice: look for water, varied landscape, and such. It returns the discussion to what is hardest: patience.

If you want to farm, location matters, yes, but that is not the same as picking a place to settle. Do something where you are, then where it is easiest to go. Do not focus all your creative energies on a place until you develop an experience level that renders you competent to tackle that decision. Delay deciding on your final settling place for as long as you can and it will come gently and easily. It will be a natural outgrowth of your farming enterprise. When you're ready, you'll know.

Being Neighborly

Farming communities are different than urban communities. To be successful as a farmer, you need to know how to incorporate yourself into the rural community, and more specifically how to interact with the farmer across the fence. Unlike in the city, chances are that the neighbor across the fence can trace his roots back, on that very spot, at least two and maybe three generations. He has watched people come and go. His daddy has told him about people who came and went. He thinks you'll come and go.

While I would hate to prejudice you about farmers, I think my general assessment is right on the money. Farmers mistrust newcomers. If that is too strong a word, then at least they are wary of newcomers — especially northerners coming into communities south of the Mason-Dixon Line. People in Dixie still fight the Civil War.

You, as the new kid on the block, need to be aware of the dynamics involved and take compensatory action. No matter how much you may disagree with the farming methods being practiced on the other side of the fence, you need to defer to the neighbor in order to be accepted.

Acceptance is important. In the city, you don't need to be accepted by your neighbors. Most people don't even know the names of the people in the adjoining apartment or townhouse. Friendship has nothing to do with geographical location. Your friends are scattered throughout the city and whether you know your actual geographical neighbor or not is no big deal.

Not so in the country. A completely different dynamic exists. You will be forced to deal with the neighbor on numerous occasions, whether it's to borrow something, retrieve a wandering calf or cut a tree in the boundary line fence. You will necessarily be doing things that affect each other. Someday you may even need a right-of-way. Rural communities have a rich tradition of neighborliness.

This is one of the tragedies, I believe, of the self-sufficiency back-to-the-land movement. People move out onto their homestead and try to do it all themselves, never realizing the joy and value of interdependence. Amish communities are perhaps the most stable communities in the world — not because they are independent, but because they are mutually dependent. They actually cultivate different talents and businesses within the community to ensure that everyone does complementary work.

If you think you are going to move out to a farmstead and do it all yourself in the old independent American spirit, you will fail. Not only is that a prideful motive, it is a brash statement. Reading through the Laura Ingalls Wilder's *Little House on the Prairie* books gives me a deep sense of how pioneers helped each other. The ones who spurned advice ended up like the infamous Donner party.

Homesteaders who gossip about their neighbors are not nearly as successful as the ones who develop friendships with the neighboring farmers, in spite of the chemicals, overgrazing and erosion. If you alienate them, or never develop a relationship, you can never reach them. But if you show genuine interest in their affairs, they will reciprocate.

Dad used to always say: "Lead by example." Don't try to convert people until you have a track record of excellence. It's easier to pull on a string than push on it. Draw your neighbors to your lifestyle by reaching out to them.

Your new community has huge resources for you, from tractors to garden planting advice. You will get ahead much faster by tapping into these resources. Here are ten principles for doing it:

(1) Don't give advice. Nothing is as unappreciated as unsolicited advice. Take your neighbor's advice without arguing and then decide what you are going to do. Few things get under the skin of a farmer more than some upstart telling him how to farm. Just keep your ideas to yourself and let the farmer "discover" them in a non-confrontational way.

(2) Watch your appearance. Most farmers are fairly conservative, church-going folks who get real uneasy around people with pierced noses. I know man judges by the outward appearance and God judges the heart, but none of us is God. People do indeed judge by outward appearance and to demand that they stop doing that is inconsistent with human nature.

I know this may anger some folks, but if I failed to mention this point I would be doing a disservice to your success as a farmer. To a straight-laced Presbyterian Republican farmer an earth muffin with nose rings is like something from outer space. The farmer doesn't understand all this stuff — especially men wearing earrings — and will naturally shun what he doesn't understand.

When you move into a farming community, you inherently bring up questions:

- What are they going to do here?
- How did they get their money?
- Can they be trusted?
- Why did they leave where they were? I haven't left for three generations.
- Who are these people, anyway?

To add to these normal questions an appearance akin to someone off the streets of a big city throws an additional hurdle into the farmers' thinking. It's just another obstacle you must overcome.

I am not talking about violating convictions to look like your neighbors. I am talking about being sensitive to others' prejudices and doing all you can to show a servant's heart, to defer to your new community. Sugar attracts a lot more flies than vinegar.

Any ostentatious exhibitions — in talk, dress or character — are inappropriate. Don't show off your education or decorate your car with offensive bumper stickers. People will find out soon enough where you stand on issues. By all means don't revel in poverty or overuse government services of any type. If you use a title, particularly one that has religious overtones like "healer" or "shaman," limit its use to circles of like-minded people. I like Allan Nation's little ditty on this one: he quotes his father as telling him that "people can only stand so much weirdness in a person. You can be a nudist. You can be a Buddhist. But you can't be a nudist Buddhist."

If you want to be thought of as a professional businessperson, honest, hardworking, trustworthy, then anything you can do to look more like an all-American boy or girl rather than a druggie at a rock concert will be helpful. Everybody: be clean and use deodorant.

(3) Offer assistance. Remember that the average age of the American farmer is now almost 60 years old — among cattle owners it's nearly 10 years more than that. In addition, most farmers do not have children helping them. The children grew up and went to town for acceptable careers long ago.

While helping out, you can learn valuable lessons. Here is a partial list of some things you may be able to help your neighbor do:

- cut firewood, including splitting, stacking, hauling
- work cattle, including sorting, castrating, vaccinating and doctoring
- build fence
- make hay, including driving a tractor or stacking hay
- plant anything
- build anything, including sheds and outbuildings
- paint
- repair buildings, including the house
- dovetail a trip to town — bring him supplies to save a trip
- haul manure

This is the stuff of neighbors' glue. Not only will you learn techniques that work, but you will also notice things that you would do differently. Mainly you will gain a friend, and to have a friend who is also a neighbor is a gift indeed.

One of the problems you will no doubt encounter is that if you practice ecological agriculture, you will not do many of the things your neighbor does. You probably will not spread fertilizer, give worm shots or pump manure slurry. But you need to look for opportunities where you can help, especially in things that you also do.

Many farmers work in town and have a real need for someone to help them in crisis situations. Perhaps the cows got out or they need to get a field of hay in during thunderstorm season. You can certainly build up IOUs for machinery work or livestock hauling by being a pinch hitter for these short-hour neighbors.

(4) Don't blab your dreams. Most farmers don't have big dreams anyway, let alone appreciating your weird ones. Tell your dreams only to people of like precious faith. That's what sustainable agriculture conferences are for.

Most farmers really do not believe you can make a living farming, especially an enjoyable and lucrative living. Sharing your dreams, then, is an affront. If it were attainable, they would have done it. The notion that you intend to do it when they did not or could not is therefore a slap in the face to them. Your dreams are actually quite threatening. Imagine being 60 years old and realizing that you have spent your whole life doing the wrong thing. Most people can't handle that so they lash back either verbally or mentally.

For a farmer to admit that your dream is attainable is to admit that he could have attained it and that has too much remorse wrapped around it to be acceptable. Don't put your neighbor in that kind of mental anguish. Be sensitive to the impoverished hard life of the farmer — his stereotype — and do not affront that. You'll gain his respect that way and be able to build a relationship.

(5) Anything you borrow, bring it back better than you found it. Be extremely careful about borrowing things, especially things you do not have much experience operating. Generally it is much better to get both the farmer and his tractor. If you break something, it hurts two ways.

First, you may have a very expensive repair bill on your hands. No matter how innocent you may have been, "it wasn't my fault" simply will not work. If it broke under your use, it was your fault, period. The only exceptions are pre-discussed weaknesses, like: "That bearing is starting to squeak. Now if it goes out, don't worry about it. I'm planning to replace it when it goes anyway." That's one thing, but an undisclosed weakness is not a credible weakness.

Second, it makes you less trustworthy to get things in the future, and this can be more devastating than the cash repair bill. The quickest way to lose help in the neighborhood is to be perceived as reckless. I know farmers that nobody around here would even loan a hammer. They'd figure out a way to break it. Your reputation takes a long time to build; guard it jealously.

If you ask the farmer to come with his equipment, you may end up paying more or bartering higher, but at least you will not be responsible for whatever happens. Generally that is worth the extra price.

Farmers are not stupid. If you borrow a hydraulic jack and leave it sitting out in the rain so that it comes back a little rusty, that farmer is going to notice. It's better to 'fess up to it up front and offer to replace it than give it back hoping he won't notice. A little crack in a sledgehammer handle is like the San Andreas fault to the hammer owner.

Be religious about this policy and you will be amazed how many things you may not have to buy. Through your own innocent negligence you may have to replace a couple of items, but that will be far cheaper than losing respect.

(6) Respect your neighbors' property. A corollary to this rule is: "Keep your animals at home." Probably the single biggest controversy among farmers has to do with boundary lines and roaming livestock.

When you wake up in the morning and see your goat contentedly chewing its cud on the neighbor's car hood, a good-natured apology won't get the job done. Believe me, speaking as a farmer, if your dog so much as steps into the field where our cows are and they run even 10 feet, that dog is dead. Period. That's the way it is in the countryside, regardless of how cute and cuddly little Fifi might be.

One of the things country folks guard religiously is privacy, and that includes control over their private property. Regardless of where you stand on the property rights issue, farmers are thin-skinned when it comes to someone trespassing or abusing their property. After all, the farmer bought it, pays the taxes on it, takes care of it, and loves it. Our society is full of people who want to dictate to others what they can and cannot do with their possessions.

One of the first principles of humanity is to leave other people's things alone. Most parents try to inculcate that idea into their children. When your cow strays into the neighbor's field, you just messed with his "stuff." When your pig roots up the neighbor's azaleas, you just messed with his "stuff." When you go see your neighbor and your lovable Border Collie tags along, only to chase some chickens around the corner or square off with the neighbor's Angus bull, that's not cute and it's not funny. You've just messed with his "stuff." Why is that so hard to learn? This is not a property rights issue; it is a moral respect for other people.

This is one of the most abused areas, and being a longtime farmer, living in a farming community, I see this as probably the single biggest point of contention between farmers and newcomers, and one of the hardest things for newcomers to "get." Farmers are generally a fairly disciplined lot. That means they have very little patience with uncontrolled animals or uncontrolled children.

When you come to the neighbor's for a visit and your small child comes along, don't let him wander out through the garden willy-nilly. It's not cute if he steps on a tomato. Shame on you for not watching more closely.

I won't belabor the point any further because I think I've made the case strong enough — "keep your hands to yourself." That includes every extension of your hands: your goat, your children, your dog and your cat.

(7) Join whatever local agriculture organization is appropriate. You probably will not join the local pesticide application group, but you may join the Forage and Grassland Council, *or* your children may join 4-H. As much as is possible, incorporate yourself into the agriculture infrastructure. Local field days are educational and social.

Most states have a forestry association that gives bus tours. Become involved with the Natural Resources Conservation Service bus tours as well. Meet and be congenial to the local extension agent and make an appearance once in awhile at the stockyard.

This includes being personable down at the farm store or feed store. When you buy bedding plants or seeds, ask the clerks questions about what people are buying and get planting tips. They have a wealth of local knowledge and will appreciate your curiosity. They will remember your face and you will become more than just another "newbie."

Enter items in the county fair and volunteer time there. These are always shorthanded and usually run by the true old-line farmers. You may join the local Ruritan club as well. Although it is not a farm organization, most of the old important farmers are members.

Help out at the local horse show or other events like firemen's parades or carnivals. Show your spirit is part of the community and you really do care about what goes on there.

(8) Invite neighbors over for a meal. You may be perceived as being a little too aggressive if you do this on a non-special occasion the first time, so take advantage of holidays and start easy with just refreshments or something.

Most people love dinner invitations. Don't wait to be invited, because you probably won't be. Farmers are reserved, you know. You be the inviter, and get those neighbors over. Fix something they will enjoy. Don't try out your turnip and rutabaga cookies on them.

This is not the time for a dissertation about diet and store-bought food. Most mainline country church groups have printed a cookbook at some time. Try to get one — the church group will appreciate your money — and look through it to see what kind of recipes people are used to in the community. This will tell you what you would expect at any church group or community potluck dinner.

Stay within accepted norms and fix plenty. Farmers like to eat. Don't encourage them to stay late, but do be hospitable and full of questions about their lives, their families and the community history. Farmers love to talk about what they remember, and most have quite long memories. Ask about "the old days" and enjoy the stories about where fences ran, where cabins sat or where the highways used to go.

Initially, stay away from any game playing. Let it be light talk and reserved "getting to know you" conversation. You will find neighbors who are gems and others with whom you may not want to develop real close friendships. Word will spread quickly that you are open and friendly, and that is a wonderful reputation.

(9) Do business in the community. Patronize your neighbors as much as possible. Buy as much as you can locally, not only from the business community, but also from farmers.

If you want to plant rhubarb, go over to the neighbors and see if they have a plant ready to divide. If you want to raise some baby calves or lambs, check with the neighbors. Chances are a nearby dairy will gladly sell you some bull calves. Most farmers would much rather sell things privately than through the stockyard. And it's much healthier for the animals.

Collaborate as much as possible. Exchange a heifer for the use of a bull. Trade eggs for milk. You'll find neighbors more than willing to haul your livestock in exchange for a Thanksgiving and Christmas turkey.

Keep your eyes open for needs you can meet. You probably will produce something your neighbor does not. See if you can enter that item into the local commerce, or barter it for services. Farmers often think of newcomers as dreamers, off chasing some fairy tale.

Doing business or barter will help dispel that notion.

(10) Talk. This is one of the hardest ones for me to do because I'm a Type A aggressive go-after-it kind of guy. I like to see the chips fly and the fire roar. But farmers love to lean on pickup trucks and talk, maybe between spits of tobacco juice. Taking time to partake of this favorite country pasttime is important (talking, not chewing).

When you go over to see the neighbor and he's propped up on a 5-gallon bucket out in the shop, you'd better just settle down from your all-fired hurry and enjoy a leisurely 30-minute conversation. It'll do you good — I know it does me.

Take time to talk. When your neighbor comes over, lean on his truck. Farmers love to see people lean on their trucks. I guess it's real close to a hug or something. Maybe it's like the cat rubbing up against your leg. Anyway, farmers get fairly lonely out there in the field and they look forward to unhurried conversation.

You'll find out about your place and the kind of crackpots who used to own it; you'll find out about the crackpots in your community; you'll find out why it's not a normal weather season and a million other things that will help you become a well-rounded member of the community.

All in all, becoming a member of the farming community is akin to joining a subculture. In fact, when I take friends down to the stockyard and they see all the old codgers sitting around dribbling their tobacco juice into soda bottles or the myriad one-gallon cans provided for the purpose, they generally exclaim: "Man, this is an entire subculture, isn't it?"

Yes, it sure is. When you realize that less than half a percent of the population is made up of farmers, you know it is definitely a subgroup. And yet it is a close-knit group with shared problems. Because farmers live with their work, their get-togethers are not like the vocational fraternity gatherings or trade association meetings. It is much more, much deeper.

To share crop failures, bountiful harvests, dead lambs and

blue ribbon heifers incorporates the deepest emotion into the vocation. Farmers do not check out of the office at 5 p.m. and walk away from their work. As a result, the commonality that bonds farmers is stronger than anything experienced by labor unions.

Add to that the belief that farmers are under siege, and you have an even deeper emotional bond. Farmers feel under siege from environmentalists, food faddists and multi-national corporations. They feel under siege from anyone who is not a farmer. That is unfortunate, but more often than not it is the case.

You are not just breaking into the local dentists' fraternity; you are breaking into a desperado camp, and desperados require you to prove yourself before they will share secrets and camaraderie with you. That is the way it is. No shortcuts.

You will find that the strength of soul and economy that you gather from farming neighbors will be unlike anything you've experienced. It will be well worth the price. And as you gain their acceptance, they will do things for you that you wouldn't believe.

Let me tick off some of the things our neighbors have done for us over the years:

- pulled a 600-lb. calf out of the swimming pool with a round bale lifter
- finished mowing our hay — several times — when our mower broke
- hauled all our livestock, and still do
- dug Dad's grave with a backhoe
- fixed our machinery — many times
- pushed in fence posts
- loaded logs with a big front end loader
- cleaned out the barn with a big front end loader
- let us board stock on a per diem basis when we had too much grass
- replaced a barn roof when it blew off
- showed me how to castrate pigs

I have been blessed with wonderful neighbors. I pray that their list would contain several items with my name beside them as well. You cannot buy the knowledge and help garnered from those special folks right across the fence from you. Yes, the ones that spray Sevin on their garden and plow the hillside to plant corn.

But you have some idiosyncrasies too — we all do. Opening up your heart to your neighbors, using these principles as guidelines, will increase the success of your farming enterprise a thousandfold.

What You Need

Although I've gone into great detail describing all the farm trappings you do NOT need, a few things are essential. What do you really need to start? Obviously this varies substantially with the scale of enterprise you plan. For the sake of discussion, let's assume we're talking about the most described operation in the wannabe world: a 10-50 acre farmstead, partly wooded, with both south and north slopes, secluded but 30 minutes from town, with a pond, stream, spring, a couple of outbuildings and livable house. Add to that the following requirements:

- Nice and quiet
- Good views
- Good schools
- Low property taxes

Any realtor who could manufacture a few hundred such places would be an instant billionaire. Now on this imaginary piece of property we're going to make a full-time white-collar living — not the first year, but within five years.

That means we'll need a combination of fast cash enterprises like pastured broilers, turkeys and eggs. We'll want to produce all our own food with enough left over to sell. We need to keep the open land grazed and that will require some fence and a few stockers, or a milk cow if we're so inclined. A couple of hogs for winter pork and multi-use lard will be on the agenda as well.

Wood heat means cutting our own firewood and being able to haul it to the house. We can get a neighbor to till the garden spot to save on equipment costs. We'll probably want to go ahead and put in some grapevines and a couple of fruit trees just to get them going as soon as possible.

Just to prove what can be done, one of our apprentices went to a 12-acre piece of leased property in North Carolina and in his first season grossed $12,000. This should inspire anyone with the potential for a viable farming income. That was not his net, of course, because he had to buy wrenches, shovels — everything.

How in the world was he able to do this? He had experience. As you look at the opening scenario, be assured that a first-season enterprise can indeed gross more than $10,000 if you know what to do.

This brings us to what I consider is the most important thing you need: experience. Lest you become frustrated because you have no experience, let me assure you that no free lunch exists. Magazines like *Countryside and Small Stock Journal* are full of articles about people who spent years with the kind of place I just described, but are still working in town and unable to make a living. Why?

Trying to build a business is hard enough, but trying to build a business you know nothing about is well nigh impossible. Everyone has to go through the trial by fire. For some reason most folks think: "I'll be smart enough to escape these crises." No, you won't. Believe me, you won't. Here are just a few lessons EVERY farmer has to learn the hard way — including me:

- Rats can carry off 50 baby chicks in a couple of hours while you sleep.
- Unseasonably wet or unseasonably dry is normal, not exceptional.
- "Normal" weather is exceptional.

- Weeds grow faster when you're not looking at them.
- The best place for a fence is where the rocks come to the surface of the ground.
- The tractor never breaks unless you're using it.
- Little apple trees have special radar that attracts deer and rabbits.
- Raccoons have a special affinity for chickens.
- No matter how good your grass is, the cows can't wait to get to the neighbor's field.
- When you leave the property, the spark goes out of the electric fence.
- When the truck is stuck in the mud, you don't get out by floating the engine valves.
- Goats think car hoods are special chaise lounges.
- To find old rusty fence wire, send out the sheep.
- Every minute of winter consumption requires 30 minutes of summer canning.
- Houses require a lot of maintenance.
- Horses eat twice as much as a cow.
- Things roam around at night.
- Only in the wild or as a hobby do rabbits proliferate easily.
- Frost damage is real and deadly.
- Deer are not beautiful animals.
- Tools really do walk away from the shop bench; they have a life of their own.

If you can't tell which of these are true and which are false, you lack experience. And more than anything else, you need experience. No shortcuts exist to getting it. I am not trying to scare you, but I am trying to help you understand that farming comes with a huge learning curve, certainly as big as anything else in life and probably bigger.

You cannot escape the learning curve. You can smooth the learning curve by apprenticing, by growing some things where you are, by working with mentors, reading, and by attending good conferences.

Although it may sound archaic, nothing beats spending time with a mentor. This old discipleship concept is still the best way to prepare for a career. Apprenticeship programs exist all over the country. Spending time with someone who knows how to do what you want to do is the best preparation you could ever get.

In addition to experience, you need connections. It takes time to build up credibility in a community. You don't just go in and start selling things the first day. You need to know where to go whom to talk with and how to get help.

Many a greenhorn has been taken advantage of by a livestock trader or equipment dealer. These folks have been in the community a long time. They can spot a novice a mile away, and if they have a thread of unrighteousness, will look for an opportunity to unload some not-so-good cattle or that tractor with the jerky hydraulic system.

Horror stories abound about someone paying top dollar for some plug ewes. Connections are how you find out who the good welder is, the honest car mechanic and the reputable livestock trader. Every community has good ones and bad ones.

This is one reason I recommend living in a locale for awhile before you actually acquire land. It gives you time to learn whom to trust. Often your best investment for the first year can be just getting to know people and asking leading questions that will steer you to the honest folks in the area. That certainly is as important as any other investment you could make in your farming enterprise.

This connection business, of course, comes with a price tag. The price is your being service-oriented. Stop off and help a neighbor get his hay in when you see him in the field. I've actually just walked across and ridden a few rounds with one, just to talk and be together. That kind of time investment will stand you in good stead, proving you are a real person, with real goals and real language.

Along with connections and experience, you need tools. Here is a list at least to get you started:

- sledge hammer — two-handed as well as one-handed
- digging iron — for both digging and prying
- wrecking bar
- adjustable wrench
- garden trowel and rake
- round-nosed shovel
- chisel and punch
- hacksaw
- pitchfork
- bench vise
- oil cans
- grease gun
- claw hammer
- pick
- posthole digger
- splitting wedges
- mattock
- hatchet

Photo 16-1. *Every farm needs a few tools: stepladder, grease gun, wrenches, sockets, heavy drill and bits, circle saw, cordless drill, bolts, screws and nails, silage fork, chainsaw, wrecking bar, come-along (hand winch), chain, sledge hammer, pick, digging iron, posthole digger and shovel.*

- nippers/wire cutters
- tape measure — 25 ft. is most versatile
- circular saw — 7¼"
- drill — 3/8" first and then ½" chuck
- heavy duty extension cords — one should be a droplight
- silage fork — for leaves and wood chips
- scoop shovel — for light fine material like sawdust
- full set of open end and boxed end wrenches
- full set of socket wrenches, both half inch and 3/8" drives
- chains — high tensile is lighter and stronger than 3/8" log chain
- pliers — channel locks, vice grips and needle nose
- screw drivers — both flat and Phillips head
- rope — braided nylon 3/8" rather than hemp, which rots
- come-along (hand winch) — at least 1,000 pounds capacity but 2,000 is better

Those are what I consider the farmstead essentials. Now comes the next group, which you would add as needed, but I want to list them to give you an idea of future expenses:

- cant hook
- welder
- acetylene torch
- chain saw
- garden tiller — preferably rear-tine multi-purpose like BCS
- level
- fence stretcher
- specialty bits for drill — ½" X 12"
- inventory of bolts, screws and nails
- cordless drill — heavy duty

Certainly a multi-year farm has more tools and equipment than this, but hold off on purchasing specialty items until you find need for them. All the above items are highly portable; they can follow you wherever you go in your farming saga.

Multi-Purpose Everything

As difficult as it may sound, you should incorporate multipurpose into every thought. It should become your password. This is, of course, a corollary principle to biodiversity. We know how important diversity is in the landscape; now we must apply this thought to buildings and machinery.

Single-use tools, machines and structures are inappropriate for a couple of reasons:

(1) They have limited application and therefore limited use. Think about a combine. This machine runs morning and night for two months out of the year and sits idle the rest of the time.

Machines should be run frequently to pay for themselves and to keep seals and gaskets functional. Because farms have a seasonal nature, single-use equipment gets used infrequently. This is especially true of tillage equipment. In general, think seriously about hiring work done that requires a specific machine, rather than buying equipment, especially if it is self-powered. Machines powered by power take-off (PTO) are not quite as bad, but still need to be used many hours per year in order to pay for themselves.

The flip side of this whole issue is a confinement livestock facility or huge factory greenhouse. They are used year-round and

Photo 17-1. *Bagging lawn clippings makes mowing the lawn a multi-purpose enterprise: generating fertilizer, herbicide and irrigation for the garden.*

Photo 17-2. *Daughter Rachel mulches head lettuce with lawn clippings. Mulching controls weeds, conserves moisture, and adds organic matter to maintain fertility.*

Photo 17-3. *Rachel checks on lawn-clipping-mulched cabbages.*

not allowed to sit idle. That brings us to the second problem.

(2) They lock you in. Since they can't be adapted easily to any other use, if you get tired of using them for what they were purchased for, that's too bad. Since you have them, you must continue to use them in order to get your investment out. Or, as in the case of huge single-use buildings, you must continue to fill them, to keep product flowing through them, in order to cash flow the bank payments.

You would be surprised how many farmers continue to make silage and plant grain when they see that it doesn't pay when compared to a grass-based livestock operation. But they have invested so much time, money, and experience in the silo and equipment that they just can't bear the thought of not using it.

Perhaps one of the most forward-thinking dairy families I know, Paul and Cyd Bickford, in Wisconsin, realized the drain that their state-of-the-art confinement dairy was putting on their finances.

They did an about-face and today, even though they have first class silos and virtually acres of concrete, they do not use them. "I can't afford to fill the silos," Paul said. I wish thousands of other farmers would be as astute.

This "locking you in" idea comes full circle when you realize what an impact it has on the next generation. Not only does it lock you in, but it also locks in the next generation. Once that poultry house goes on that farm, it has to be used. Perhaps the next generation doesn't want to grow chickens. Maybe a son or daughter wants to grow orchard fruits or blueberries. The house, though, requires such an investment in time and money that the other options can never be explored. You can't just sell the house and get your money out of it.

The same is true with big machinery. Since you can never sell it for what you paid for it, you feel economically and emotionally compelled to continue using it whether you enjoy it or not. And this same bondage passes on to the next generation.

Just to show you what a lunatic I am, I think this principle even applies to a house. Why in the world would someone make a house that will stand for 100 years? One of the most interesting housing stories I ever read was in an early issue of *Mother Earth News* about a family who built a huge 30-ft. diameter, 30-ft. high teepee. The title of the story was "Kon Tipi."

The basic idea was that the two-story structure cost only $10,000 and half of that was in the high-tech canvas skin. It was a special fabric that would last for 10 years. Even though the family had to rebuild their house every 10 years, the housing cost was only $1,000 per year, and each time they could modify it to suit their changing needs. Most mortgages are that much in two or three months, let alone a year. These folks could spend cash on their house, stay debt free, and save a little along to be ready to replace it when necessary.

The story made a huge impression on me as my friends began building houses. Invariably, I'll go over and see this brand, spanking new house to ooooh and aaaaah over it. The furniture isn't even moved in it. Within five minutes of the tour, the owner says: "Now if I had it to do over again, I'd move that wall over a little and switch that cabinet around. See how that door swings the wrong way?"

I'm incredulous. No one has even slept a night in the house, and it's already not what they want. Why saddle ourselves with a lifetime of debt, and saddle the next generation with our mistakes, by building houses that will last a century? Should any of us be so arrogant as to assume that people a century from now will be so obtuse that they will not be able to make major improvements on what we've designed?

We need to keep the door of innovation, creativity and refinement swinging freely for the next generation. Single-use, highly specialized buildings and machinery are a sure way to box all of us into something that could be obsolete next year.

Whenever you build something or buy something, ask yourself these questions:

(1) How much will this strap the farm economically? If you can't abandon it in a couple of years without seriously affecting the balance sheet, it probably is not a good investment. That may sound like a harsh statement, but I've seen too many farms held hostage, economically and emotionally, to a huge capital expenditure. This bondage can last for decades and even into the next generation.

(2) What else can I use it for if my current use plans change? If the cost of retrofitting for a different use will be as much as a new structure, think real hard about proceeding with your plans.

(3) What else can I incorporate into this structure besides the basic use? Usually a building can be used for more than one thing. Although normally there is an overriding need that makes you want to build something, try to incorporate as many other uses as you can. For farm structures, the main cost is in the building's shell, so the more you can add inside, the better off you'll be.

Often adding a few feet to the height can make room for a mezzanine or partial second floor. The additional cost is negligible, but the added square footage is double for every two-story portion. This is why two-story poultry houses used to be common.

(4) Emotionally, could I bring myself to rip this building apart next year and start over with something more functional? If I don't want the building here, could I move it easily? The whole issue of flexibility is important. If you're learning you'll be modifying. When buying machinery, think of its multiplicity of function rather than just price. For example, four-wheel drive tractors are now becoming quite popular. The difference in cost between a two-wheel drive tractor and a four-wheel drive is relatively small compared to the overall price. If you add a front-end loader to the four-wheel drive, you have an extremely versatile machine. We have two tractors, and both of them are four-wheel drives with front-end loaders. A front-end loader can dig, move material, lift and push. A two-wheel drive tractor is virtually worthless with a front-end loader because as soon as you lift up on the bucket, the rear wheels lose traction. It is definitely worth the extra money to get a four-wheel drive that will allow you to get the value out of your front-end loader. Buying a front-end loader just to watch the tires spin is not my idea of cheap entertainment. We have a set of forks that mounts on the same points. In a matter of minutes, we can remove the loader and put on forks for lifting pallets and logs. You can purchase all sorts of implements for the 3-point hitch, like backhoes, log winches, chippers or whatever.

This is the secret to the BCS garden tiller. What you buy is a power plant and then add attachments according to need. Maintaining engines is far more expensive than maintaining attachments. Put your money in a good basic engine and then hang things around it as needed.

I like pole structures for their versatility inside. The poles are strong enough to bear weight either vertically or horizontally. The whole building requires less lumber than post and beam or stick structures. I am partial to wood as a building material because it is a renewable resource.

I like buildings built on skids. We have built several 12-foot x 20-foot buildings on 20-foot locust poles. This is a fairly large building, but can be towed with a small tractor if you tire of where it sits. They are multi-functional by being big enough for animals, yet small enough for storage buildings.

Here is a list of single-use machinery that I would caution you about:

Combine	Log Skidder
Plow	Disk
Corn planter	Small grain drill
Silage blower	Self-unloading silage wagon
Corn picker	Skidsteer loader
Mower/conditioner	Liquid manure spreader
Potato picker	Vegetable pickers

Here is a list of multiple-use machinery:

Dump truck	Dump trailer
Backhoe	BCS garden tractor
Four-wheel drive tractor/loader	PTO Manure spreader

While these are not by any means exhaustive lists, I think you get the picture. Obviously, some single-purpose equipment is necessary depending on what you want to do. For example, we have a scalder/dunker and an automatic feather picker for poultry processing. You can't do too many other things with those machines, but the volume of income they generate makes them an incredible investment.

If you are going to produce ten acres of vegetables, perhaps a diesel Goldoni is your machine of choice, rather than a smaller BCS garden tractor. Keep your machines scale-appropriate. As a farmer who practices diversity, direct marketing and size-appropriate policies, many of the industrial farm machines will be inapplicable and unnecessary in your operation.

Just remember that you cannot unbuy something. Be honest about how supposedly necessary it is and then think through your options. You don't want to be stuck next year saying "oops" about a major purchase. Finding good tools and building functional facilities is an enjoyable challenge if you keep focusing on multi-purpose everything.

Where to Buy Things

As you begin outfitting and maintaining your farming enterprise, you need wisdom regarding where to purchase different items. Some things can be acquired through a wide variety of venues and other items will offer you little choice as to where to buy. Let's get acquainted with some of the options.

AUCTIONS. You can learn about these by watching the classified ad section of the local newspaper or farm press tabloids. Seldom are these publicized in any glossy publication. Although they can be held on any day of the week, they are usually on Saturday, with a time slot reserved for big-ticket items like farm machinery or real estate.

Scan the items listed in the advertisement and see if the sale has something you need. Don't waste your time going just because it looks interesting. Auctions can be addictive sources of recreation: don't let them be. Most auctions are estate sales or liquidations due to financial changes.

Like every other market venue, auctions have both good points and bad points. The overriding negative is that you buy it **AS IS**. In other words, items come without any guarantee (unless something is still under warranty, in which case the auctioneer will tell you) and they stay right where they are until you move them.

If you buy a shovel at 10 a.m. and come back to where it was and someone has walked off with it, you just bought a shovel you'll never use. Don't let items out of your sight. Move everything you buy as soon as possible — the day of the sale at least.

Generally auctioneers are truthful about items, sharing whatever they may know. If it's a machine, the big question is, "does it work?" Unless you are extremely knowledgeable about a certain item, never buy something that will not work. Don't believe that "all it needs is a little gizmo here and it will run like a top." Forget it.

Watch the light turn on, the motor spin, whatever. Just remember that at auctions, generally nobody guarantees anything. When that auctioneer points at you and says "SOLD!" you just bought it. Period. This policy means that, all things being equal, auctions present more bargains than anyplace else. But all things are usually not equal. And that's where the fun begins.

Let me tell you what I do at auctions. After getting my bidder number, I go look at the stuff. Almost always a couple of older men will be sitting on a piece of equipment, kind of away from the mob, chatting. Whether I know them or not, I kind of join them. I say "kind of" because this is the art of getting information without being suckered.

Farmers love to joke with greenhorns. You don't want to appear like you don't know anything, so absolutely do not walk up brusquely. Amble up as if this group is as good as any other around, and do NOT say:

- "Hi, I'm new around here. Can you tell me what that is?" *They'll have a great time telling you a bunch of lies. Oh, they don't mean to hurt you. They'll have a twinkle in their eye and a smile on their face while they feed you a bunch of hogwash.*

- "Can one of you gentlemen help me? I'm looking for some information." *Way too aggressive and naïve. You're coming on too strong and announcing how ignorant you are. Don't be straightforward.*

- "I just bought a farm." *They will think you've got a pocket full of money and they would enjoy watching you be taken down a notch or two. They think you are one of those city folks who is driving land prices up and squeezing them out of business.*

- "Do any of you work here?" *Any employee on that farm is bustling around cleaning out sheds and toting stuff out to the yard. Any farm employees will be somewhere right around the center of activity — probably up by the auctioneer.*

- "What are you all going to buy?" *This is threatening these guys because if they have their eye on something, they sure don't want you to know about it. Farmers get real possessive about things at auctions and they don't like people coming around the things they plan to bid on. You'll never get a straight answer to this question.*

What I do is come up close, but not right into the circle. I leave them some space, but let them know I'm listening and will probably eventually join the conversation. I laugh with their jokes and nod vigorously in agreement when they complain about environmentalists or multi-national corporations. I let them know by body language that I am more similar than dissimilar.

After a few minutes, I'll offer a little tidbit into the conversation. Finally, I introduce myself and just continue in conversation. This whole process may take 20 minutes. Once I get to the point that a passerby would identify me with this group (even though my cheek isn't puffed out with a wad of tobacco) then I'm ready to move in for some information by simply steering the conversation in my direction.

ME: "Well, it's amazing this place got to this point." *This is a benign statement. I may not know if the owner died or went bankrupt or whatever. The other men have no way of knowing — and have no reason to think I'm pulling a fast one — that I really don't know anything. But they will surely have something to say about this statement, and it will be my lead-in to getting information.*

THEM: *(laughing)* "Yeah, remember the day Bill lost that tractor tire pulling the silage wagon down the road for Mel? It's a wonder he hung on as long as he did." *Now I've got some information. Apparently the former farmer—whether he was owner or manager, I have no way of knowing yet—was rough on his equipment. That will go a long way in helping me formulate my opinions about the value of things.*

ME: *(being adventuresome)* " Well, at least he didn't get hurt." *This is fairly safe. If he'd been hurt, it would not have been funny. And the fact that he hung on indicates he continued to run things. All I'm trying to do is string along the conversation without showing too much ignorance.*

THEM: "No, but he was a reckless son of a gun. That's why they had to rebuild the front end of that tractor three times. He was always using it like a bulldozer." *It so happens that the four-wheel drive tractor was the item I was interested in. Now I know the tractor has been abused and the front universal joints have been replaced several times.*

I could drag this on, but you get the drift. Engage anyone who looks willing to talk in conversation and find out everything you can about the farm, about the way things were cared for and about the circumstances surrounding the sale itself. Sometimes the sale occurs after things have sat for some time. Eavesdrop on every conversation you can to learn as much as possible about the farm.

Often you will find things at an auction that you can't find anyplace else. I've bought buckets of old mattock heads for a dollar. All they need are handles and we're in business. We bought our manure spreader at an auction, and a hay mower we used for several years.

Auction psychology is a whole science, but here are a couple tidbits you may want to remember. In general, wait until somebody else bids. Never be the first bidder. If you really want an item and the bidding is tapering off, yell out a new dollar amount that is a significant step above where the bidding is lulling. Often this psyches out the other bidders.

Set a ceiling and stick with it. You can be swept along easily in the euphoria of the moment and overspend. If the bidding starts where you hoped to end as high bidder, you've just been outclassed.

Remember that many people at the auction have inside information. If you don't have inside information, be cautious. Every sale burns someone. Buy based on knowledge and need. Sometimes items will go within a fraction of brand new ones. But sometimes you will get incredible bargains.

Auctions are like anything else: the more you attend, the more educated you become. I encourage any would-be farmer to attend several auctions a year just to keep up with what's going on and what kind of people are buying. Don't feel like you have to buy something. I've left empty-handed more times than I've purchased something. But I think the purchases I have made have been great bargains.

Once you go to a few auctions, you will develop a better sense for what you may find there, including what it means when the phrase "and many items too numerous to mention" is tacked on at the end of the ad. If you really believe the item you are looking for will be at an auction, keep going. I've spent a couple of years going to auctions to buy a specific item. You may go to five and that item consistently sells for a high price. Then one day you go to one and for some reason nobody is interested in it. Suddenly your patience pays off and you make a bargain.

And remember that no matter how much you think you know you can never be sure you've got an auction figured out. I've gone to auctions in terrible weather thinking nobody else would be there, only to find out everybody else in the county thought the same thing and there were more people there than would have been on a pretty day. The same holds true for a real pretty day.

I've heard old-timers say: "Never throw anything away." Once you've been to an auction, you'll see why. Virtually everything has a value. Many years ago I had gone to some auctions and watched chicken crates sell for a dollar. When I really got into chickens and needed some, I decided to go and pick up some bargains. Little did I know that the new rage in upscale interior decorating was the lowly chicken crate. I watched incredulously as manure-covered, broken-down chicken crates sold for $25 apiece.

I could scarcely believe my eyes when shortly thereafter on one of my annual excursions to the mall I saw top-of-the-line clothing displays perched atop chicken crates — in a prestigious department store! Who can figure?

CLASSIFIEDS. Every local newspaper has a classified section that should be perused several days a week. I look at the classifieds routinely because I've found items there that may lead to something else. For example, someone selling "butchering hogs" may have a couple of sows and be able to provide me with weaker pigs.

Someone liquidating a rabbit business may want to get rid of the cages, even though the cages are not advertised. Someone selling pick-your-own bramble fruits may be a good connection for getting some healthy canes cheap for your own raspberries.

Of course, you have to stay up on the culture to make them meaningful. Our family is conservative and would almost qualify as Amish in many lifestyle decisions. Although we do not wear Amish clothing, we do purchase things from the thrift store. Daniel and I wear work shirts from the local uniform cleaning service, which stockpiles used uniform shirts from area businesses. They don't cut off the name tags and it's quite a joke around here to see who Daniel and

I are for the day: Rodney, Jack, Dale, Cook.

It's especially interesting when I wear a shirt from a poultry conglomerate. When I go to the farm store to buy something, the clerk will ask me politely about my chicken houses and I tell her everything is just fine. I walked up to a counter at a hardware store one time and the clerk, being extremely perceptive and friendly, looked right at me and said: "Hi, Charles, what can I do for you today?" I looked behind me to see whom he was talking to. He must have thought I was a loony-bird, not knowing my own name. Then I realized that I was wearing one of these uniform shirts and tried to recover good-naturedly: "Oh, hi, yes, why sure, uh, uh, I need two sheets of that translucent roofing."

Our being "out of touch" perhaps climaxed one evening when I was reading the classified section. Daniel had been discussing tanning some rabbit hides and Rachel was talking about sewing them into gloves and slippers if he could figure out how to do the tanning process. We had gotten a couple of books on the subject and it was an ongoing point of discussion. Here I was just scanning the classified section and suddenly it leaped out at me: "Tanning Beds."

I jumped up and exclaimed to Teresa: "Hey, look at this. Here's someone that might be able to get these rabbit hides done. I wonder how it works. Hmmm. Tanning beds. Amazing."

Anyway, classified ads in the local newspaper can be entertaining as well as helpful in locating items you may need.

TRADER BULLETINS. Most communities now have some sort of trader bulletin that you buy at filling stations or rural hardware stores. We have found these quite effective for both buying and selling.

The big advantage they have over newspaper classifieds is that they target people who specifically want to read them. Since they do not come free, only serious lookers buy them and read the ads. In addition, in our area, these carry probably 10 times as many ads as the newspaper classified section.

While we by no means pick one up every week, we do get one now and then just to look it over. But when we are really looking for something, we get one as soon as they come out. Good deals routinely last less than a day before they are purchased.

The advantage to classifieds is that you can go look over the item and talk with the seller. At auctions this is more difficult and you have less time to cogitate.

If you are looking for something specific, this is an excellent way to find items. The ads are often free, as opposed to newspaper classifieds, which are not. These also carry reasonable display ads that can give you good exposure without the cost of a newspaper ad, kind of a first publicity opportunity.

CATALOGUES. Becoming more and more popular, the catalogue venue allows you to make purchases without leaving home. Emotionally, a catalogue offers the advantage of keeping you more focused on the items you need and less apt to make impulse purchases.

When you walk through a store you can easily get caught up with all the merchandise enticing you to purchase. In the quietness and reason of your home, however, all those multi-sensory stimuli do not exist and you can stay on track easier.

With the efficient transportation system we have now, as well as easy over-the-phone payment abilities, shopping by catalogue is a wonderful option and one I'm sure will continue to grow over the next few years. Rather than hundreds of people driving to stores, all these items can be shipped using vehicles that will make the trip anyway.

Most hardware and farm type stores are not open after daylight hours. As a farmer, daylight hours need to be spent outside working. When you start taking those precious hours and devote them to shopping, you'll chalk up some big losses on your work schedule.

Another reason I like catalogues is that they give me ideas. Often the explanatory notes give us ideas for things we can either duplicate in the shop or adapt to our needs. To tap into all the creative thinking that goes into product development is a real asset, even if you end up making instead of buying the item.

Photo 18-1. *A healthy assortment of catalogues provides ideas and products for your farm.*

We receive several catalogues, with various specialties. As far as I'm concerned, you can hardly receive too many catalogues. These outfits inventory items that cannot be found within a thousand miles, thereby increasing the options we have.

HARDWARE STORE. Supplies you need may be purchased at the hardware store, especially consumable type things like plywood, paint or shelving brackets. You will probably buy most household items like plumbing parts or light fixtures at the normal suburban hardware store.

Large volumes and specialties make these venues desirable for generic type things. If you are fortunate enough to still have one around, old-time hardware stores often have unusual items or non-standard parts that you can't find at bigger, newer stores. Generally too these neighborhood hardware stores stock in bulk bins. Instead of a box of bar-coded plastic baggies, you can actually just bring home the merchandise.

We are blessed with one in our neighborhood and I found the old-style Oregon chainsaw files hanging on the tool board in the back room. This style, with a horizontal lip that slips between the chain and bar, is far superior to the newer types. I bought enough to last my lifetime. Now all our apprentices want one but to my knowledge no one makes this good kind anymore. Are any manufacturers out there listening?

FARM STORE. I am making this a different classification than the hardware store, although many hardware items like nuts, bolt, nails and tools will also be carried in a typical farm store. When I think of farm store, I think primarily of the local outfit that specializes in agricultural items like chicken feeders, cattle ear tags and halters.

In many areas these will also be a cooperative, which you may join. This entitles you to a share of the dividend distribution based on your purchases throughout the year. I have found quite a bit of discrepancy in prices on generic type things between the farm store and the hardware store.

For example, we buy quite a lot of ¾" plastic pipefittings for our watering system. Each fitting must have a pipe clamp on the ends to hold the polyethylene black pipe. At the hardware store these pipe clamps are $1.32 apiece and we use probably 100 of these per year, either in maintenance or new water lines. At the farm store, these clamps are only 75 cents apiece. On the other hand, the plastic fittings are $1.39 apiece at the farm store and are only 25 cents at the hardware store. How can you figure that? These two stores are only two miles apart. Many items, of course, are within a fraction of each other, but be aware that some have large differences.

I must confess that I wrestle with the need to spend our dollars here locally or let them go to a catalogue outfit hundreds of miles away. I'll let you be the judge on how you balance those loyalties. Generally, I purchase what I can at the local farm store because this helps maintain our agriculture infrastructure. We are blessed with a couple of extremely good farm stores and I would hate to see them close.

We farmers should support the local infrastructure, in my opinion, in order to keep our purchase prices low and availability local. I would hate to purchase everything from catalogues because often when I want it I really need it — NOW. Be aware that every dollar spent locally churns back through the economy to keep neighbors' businesses healthy. While I do advocate price comparing, I am not a "shopper." Often you can spend an hour and $2 worth of gas trying to save a dime. I'd rather keep the place open where I have a relationship than take all those years of good, local service and throw them away just for a few bucks. By the way, you will be hard pressed to find me inside a Wal-Mart.

In general, keep your buying options open. The more of these venues you patronize, the more knowledgeable you will be about product choices and prices. Don't let yourself get too tied down to one vendor just because that's where the local farmers spend most of their money. Diversify your acquisition portfolio and you will be better positioned to make wise choices.

Searching for Answers

Where do you find information? The success or failure of any enterprise often hinges on how accurate, or how good, the information is that comes to the person making decisions. Information and decision-making go hand-in-hand. If you are going to succeed in your agricultural enterprise, you must carefully discern the sources of your information because information is not objective.

Data are highly subjective; the philosophical underpinnings of the information-gathering person, agency or organization bear on the findings. Anyone who does research knows the importance of asking the right questions. If we ask the wrong questions, fail to ask the right questions, or fail to ask enough questions, our research will be skewed.

To show how important this is, look at the philosophy of conventional agriculture: bigger is better. The supermarket chain, Farmer Jack, perhaps best illustrates the average farmer's thinking with their slogan, prominently displayed on shopping bags: "Pile it High. Sell it Cheap."

Dramatically absent is anything about quality, about nutrition, about cleanliness, about quality of life for those who produced and processed the food. It's just assumed that more is better, cheaper is better, thank you. Food is viewed as just so much material we must ingest every day.

The same notion comes every day from conventional agriculture experts when they talk about "plant food." Their view is that the soil is just so much material holding the plant up, and we must dump on materials to "feed the plant." Nature feeds the plant through the soil, through the billions of microorganisms like *Gibberella* and actinomycetes, earthworms and azotobacter aerobes.

As I mentioned in the chapter on philosophy, our Western mindset is mechanistic and in this area we have excelled. But when that mechanistic mindset is superimposed on the biological world, it completely jaundices our thinking. Instead of a plant or an animal being a responsive (albeit instinctively) living thing, it is considered by some to be just so much protoplasm composed of so many protons, neutrons and electrons.

Western livestock research fails to ask animals anything. We introduce new species of grasses or grains and never ask the animals which ones they prefer. All we're interested in is volume; it doesn't matter that the animals don't like it and do poorly enough on it to require medications or synthetic supplements.

Classic organic production research during the 1960s and 1970s at America's land grant colleges revealed the importance of philosophy. The Ph.D.s (I call them Post Hole Diggers) took test plots that for decades had been subjected to all sorts of strange compounds – including DDT — and planted corn. On some, they did not add anything; those they called organic. On others, they added chemical fertilizer, pesticides, herbicides — the whole nine yards; those they called the conventional chemically fertilized plots.

At the end of the season, they measured the production from the two plots and found that the "organic" ones did not do well. I wonder why? Anyone who knows a lick about biological systems knows that living things (the soil is a living thing) do not recuperate overnight from years of abuse. The supposed scientific research was asinine. And yet these conclusions were replicated and promulgated all around the world to prove the unacceptability of organic farming.

It fueled the notion like that espoused by the Hudson Institute that non-petroleum farming cannot produce enough food on a small enough acreage.

This research skewing, whether contrived or due to ignorance, is nothing new. André Voisin encountered the same thing when he was formulating principles for effective grazing management in France decades ago. Much of his *Grass Productivity* book deals with the skewed findings of contemporary researchers who wanted cookbook recipes instead of letting natural principles speak. Certainly Allan Savory's work in holistic management has met with similar opposition from researchers with preconceived antipathy toward any truly novel idea.

But beyond asking the wrong questions, it is important that we ask the right questions. Since our paradigms define the limits of our questions, our paradigms also define the limits of our creativity. For example, while the organic community runs around trying to figure out how to have an organic cattle feedlot, I ask: "Why have a feedlot? Why feed ruminants grain at all?"

While the sustainable ag community is tripping over itself with Integrated Pest Management to reduce pesticide applications on corn, I ask: "Why grow corn?" Roughly 70 percent of all the grain grown in this country goes through multi-stomached animals, which God created to eat forages instead of grains. Imagine what it would do to our chemical usage, our erosion, indeed our culture, if 70 percent of the land currently growing row crops were converted to perennial prairie polycultures. Then the grain could go for human consumption, swine and poultry.

Most of the "problems" we spend millions of dollars trying to fix would become non-problems. They would simply cease to exist. I am amazed at how much money we spend on things that — if the models were right — would cease to occupy any of our time.

Supermarkets have now taken on the tag line of "and Drug" as our devalued food supply comes full circle. If all the time, energy, money, and brainpower put into the pharmaceutical industry were devoted instead to creative animal production in such healthy models that medications were unnecessary, we would be light-years ahead of where we are. Instead, because we view animals as machines to manipulate any way we see fit, we crowd them up in atrocious husbandry models, feed them junk, shoot the medication and synthetic supplements into them, and then institute regulations to protect us from *E. coli* and *Salmonella*, as well as from manure lagoon blowouts and stinky air. Then we spend untold dollars dealing with food borne illnesses and doctor visits. Wouldn't it really be easier just to get the model right?

One of the most articulate examples of this is in John Ralston Saul's book, *Voltaire's Bastards — The Dictatorship of Reason in the West.* Describing the French schools that train government workers, he writes: "In a sense the training in all these schools is designed to develop not a talent for solving problems but a method for recognizing the solutions which will satisfy the system."

To really question current paradigms would threaten the flow of corporate grant money into the land grant colleges. A few years ago, after listening for the umpteenth time to the ag experts pontificate about injecting antibiotics into the conjunctiva of cow eyelids to fight pinkeye, I began a one-man campaign to educate these academics. We use kelp (seaweed) as a natural mineral supplement and haven't treated a case of pinkeye in 20 years. This experience has been duplicated by neighbors.

Soon the extension agent, the extension livestock specialists, the agronomists, the forage specialists — they all came, took copious notes, slapped me on the back and said I was doing a great job. BUT, they couldn't do any research to verify my experiences without some seed money. And they certainly could not say a peep about my experiences because they were just anecdotal and not double-blind research.

The next month the radio crackled with more pinkeye updates, all spouting the same stupidity, and it remains so today. If they told anybody what we were doing, it would destroy their precious "seed money" from the multinational corporations. Money talks.

If you think for a minute that anyone who works for the USDA *[I like Bud Kerr's definition — after two decades as head of the USDA's Small Farm branch, he called it the US Department of Aggravation]* is an impartial expert, forget it. The system has an entire hierarchy to stop any major leaks in the dam that holds back the truth. Every now and then a great extension agent or USDA official comes around, but they are as scarce as hen's teeth. Most burn out quickly as they tire of fighting the system, or compromise their message enough to quit being a threat to the precious system.

The reason this whole discussion is critical for you is because time and time again I've consulted for farmers who have received advice from these people that destroys their enterprise. Couple that with the fact that recent surveys show that 87 percent of farmers make their decisions based on what a salesman says, and you can see why farmers go belly up.

I've listened to folks tell me that the USDA guy — doesn't matter what branch he's with — recommends applying two tons of this or that when the farmer doesn't even have an animal on the farm. Why would anyone need to grow more grass if he doesn't have a cow to eat it?

You can go to any farm show in the country, and the exhibitors, as well as the speakers, will be selling something. Always beware of people selling things. That's why I don't sell anything — except our farm produce and an occasional book. I don't want to compromise my integrity by selling any of the products I endorse.

Of course the veterinarian is going to tell you to use some medications, or vaccinations — she gets frequent visits from pharmaceutical reps, explaining how to use this or that, leaving off note pads and pens, baseball caps and tickets to the football game, passes to the local club and who knows what else. This is standard fare in doctors' offices as well. Why do you think most doctors oppose chiropractors, homeopathy, and nutritional therapy?

While I do not believe the average vet wants your animals sick, you must understand that he is coming from a mindset, or worldview, that is fed from the university system described above and is nurtured at the corporate nipple — or wined and dined, whichever picture you like. Through no malicious intent, the vet's advice is necessarily jaundiced away from personal can-do methodology and is predisposed to offering things with which he's familiar — shots, pills and syrups.

If you ask the local feed mill manager how to raise baby calves, for example, he will look over his arsenal of grain and give you a formula that is guaranteed to make him money. Think about it — how many industries thrive and survive based on how much money they can get farmers to spend? How many cities thrive based on the amount of wealth they can extricate from the countryside?

As Wendell Berry so eloquently says: "What's wrong with us generates more GNP than what is right with us." Healthy, happy people don't need to go to psychiatrists, doctors and detention centers. Contented folks don't need pills and alcohol. Healthy, productive animals and plants don't need pharmaceuticals, vaccines and syrups. They don't need chemicals, genetic engineering, and toxic waste experts.

One of the most important things for you to understand is that our whole agricultural system stays alive as farmers die. If you chart the USDA budget growth, it is inversely related to the number of farms. The USDA budget now exceeds the entire farmgate value of all agricultural products in the U.S. As USDA funding increased, farmers decreased by millions. Charles Walters, editor of *Acres USA* magazine, queries: why would anyone follow the advice of an organization devoted to annihilating its constituents?

Innovative information inherently cannot come from the government. Allan Nation, editor of *Stockman Grass Farmer,* describes this phenomenon as the majority rule. Government advice must be acceptable to 51 percent of the population — that's how our system works. But you and I both know that true radical information will never be acceptable to the mainstream. Every great thinker, every great innovation, has been laughed at and scorned by the elite of the day.

Eighty years ago all the U.S. generals agreed that these new flying machines would never have a significant military function.

History is full of these tidbits, showing the inability of the mainstream to grasp the impact of changes that were going on around them.

Especially when you are a beginning farmer, these well-meaning advisors will swarm around you with sincerity. They desperately want you to succeed. They have your best interests at heart. They are more interested in your farm than you are.

Just remember, there's no free lunch. Be extremely wary of any "free" advice, whether it's from someone selling a product or service, or from one of those public servants down at the extension or conservation office. They are products of a system carefully orchestrated over the years to maintain an agribusiness infrastructure of pharmaceutical companies, fertilizer dealers, equipment distributors and auction barns.

Just as you would validate a decision with a "second opinion" in medicine, get multiple opinions about your farm. Be eclectic. You absolutely cannot get too much information.

I am familiar with many things: biodynamics, holistic management, permaculture, remineralizers, pyramid practitioners, conventional organics, conventional feed and forestry practitioners. I haven't met a person yet from whom I couldn't learn something.

One of the biggest pitfalls in a "movement" is the temptation to get cultish about it and refuse to realize that it is only part of the picture.

For example, one of the problems in holistic management, which is a thought-process model, is that it cannot be more creative than your own exposure will allow. Just because you come up with the best alternative out of the several you subject to analysis does not ensure that you've come up with the best solution. The best option may be one with which you are not familiar. Rather than guaranteeing success, therefore, the decision-making may spell failure. The best way to make sure we have all the options on the table is to make sure we expose ourselves to as many options as possible. This in no way impugns any thought model or movement; it only shows the limitations of each one's contributions.

This is where I have such a big problem with the USDA. If these folks would put some anecdotal information out there — with a caveat, if necessary — to expose people to more than things from the experiment stations, our decision-making could benefit from a greater number of alternatives. But these government specialists hardly give the time of day to those of us whose experiences run contrary or in addition to the mainline thinking. They present the paper, or the bulletin, as if this is the latest and best and it denies all the backyard experimentation, all the counter-systemic alternatives being happily practiced throughout the world.

Some rays of sunshine have appeared on the horizon, though, because now some individuals within the USDA organization are pushing in a different direction. Enough of them now exist that they can't be hushed fast enough. New sustainable agriculture programs have done some helpful things. Of course, I would not admit that these positive developments are occurring only because of the USDA.

I think progressive-minded people figure out a way to do things completely independently of government programs. What has happened now is that some funding has been released that those of us going in a different direction can tap into. We would have gone the different direction just as effectively anyway.

It's kind of like listing all the modern products we owe to the publicly funded space program. Who is to say we wouldn't have had them anyway? Perhaps they would not have come as fast or perhaps they would be more expensive. But to suggest that the only reason we have sustainable ag research and the only reason we have high tech electronic gadgetry is because the government paid for the research is patently ludicrous.

Anyway, the rule still holds. Check out the philosophical underpinnings of the research before you accept it. If the folks articulate a good philosophical foundation in the preamble of the USDA research bulletin, then it probably contains some good information. The information is only as good as the mindset of the authors, no matter what the organization.

Although I could be accused of blasting the USDA, certainly plenty of private information sources are as bad or worse. Slick magazines carrying advertisements for the agri-industrial complex certainly cannot be trusted, nor can think tanks with a decidedly big business bent, like the Hudson Institute headquartered in Indianapolis.

If you want to be a successful farmer, the best place to go for information is to successful practitioners. Stay away from the guy who complains that "there ain't no money in farmin.'" He obviously doesn't have the information you want.

Read books and magazines that are can-do, upbeat, and alternative to the status quo. Your success or failure will depend largely on how willing you are to depart from "accepted practice", as in Robert Frost's poem, *The Road Not Taken:*

> Two roads diverged in a wood, and I —
> I took the one less traveled by,
> And that has made all the difference.

Watch the advertisements in periodicals — they will tell you a lot about the philosophy and the thrust of the magazine. If the ads are for products sold by the conventional system, beware. If you truly are seeking truth, you will have a gut feeling about the right thing when you find it.

Attend alternative and sustainable agriculture farm conferences. These are as important for the information they present as they are for the relationships you'll cultivate. Look for conferences that have few if any university and government speakers, but rather concentrate on farmer talks and farmer panels. While there, go out of your way to talk to other attendees, the speakers and exhibitors. Develop friendships, resources and contacts. Some of the best ones going are the *Acres USA, Stockman Grass Farmer,* and *Small Farm Today* conferences.

Visit successful farmers. Be careful about interrupting them — remember, the extension service has conditioned people to expect free farming advice. Look where that's gotten us. One of the

Photo 19-1. *Your own personal library ranks high on the list of information sources.*

Photo 19-2. *You can't overexpose yourself to good ideas. Magazines help you keep abreast of the latest, greatest models.*

saddest occasions in my life was visiting a highly successful swine producer who had just won some outstanding awards from the Farm Bureau Federation, that pinnacle of conventional wisdom. But in the solitude of his living room, he confessed that he was hopelessly in debt, had many sick animals and would not farm if he had it to do over again.

It was a moving afternoon for me because I had not yet left my outside job to come home and farm full-time, and I wanted to make that leap just as soon as possible. Over the next few months, as I saw foreclosure hit this farmer friend, watched his dreams vanish before his eyes, watched the pressure take its toll on his family, he took me aside one day and said: "You know what really hurts? I did everything – everything — just the way the experts told me to."

Any government agent who reads these words should be stricken to the quick, should be sobered, by this statement. The emotional carcasses in America's rural communities, left to rot by supposed farming experts, are monuments to arrogance, money and power. We are awash in bland and unsatisfying vegetables, potentially toxic meat and devitalized food — this is the legacy of a wrong philosophy skillfully implemented and nurtured at the public trough. Heaven help us.

You can be a part of the solution by opting out, by taking a different path, by refusing to be "taken in." Visit farmers you respect, pay them for their time, trade work with them, and be willing to learn. Successful farmers are out there; ferret them out. If you search for them, you will find them and be blessed with wisdom beyond anything you could glean from mainline books or periodicals.

Apprenticeships are fast becoming a source of agricultural learning. Vegetable producers especially utilize apprentices because of the high summer workload. You will have the advantage of experience with your education. Show yourself to be a willing worker, an astute observer, and you will find many farmers happy to encourage you.

Most of us alternative farmers are overjoyed to see other folks catch the vision and begin farming. Most wannabes have no idea how much they don't know. You can't imagine the amount of information necessary to become a successful farmer — from how to hold a post hole digger and shovel to how to manage grass. Mechanics, landscaping, agronomy, animal husbandry, plant genetics, business finances, marketing — a whole host of things is better "caught" than "taught."

If you can see a master implementing these things, and work alongside him, you would be surprised how much you will pick up in exchange for some labor. Most farms are notoriously short of labor and most wannabes are notoriously short of good information. Most farmers can't pay white-collar wages and most students can't get good information. By wedding the two needs, the farmer gets affordable labor and the student gets good information. Seems like a good trade to me.

The bottom line for getting information, then, is to examine the philosophical base first and make sure it lines up with your thinking, beware of free information, read, read, read and personally acquaint yourself with the kind of farmer you aspire to be.

Brainstorming

Brainstorming sessions are crucial for two things: problem solving and capturing our full potential. While these may seem the same, they really are not. Recognizing something as a problem and then trying to solve it is completely different than trying to fully capture our potential. If we never want to go more than 20 miles per hour, we never have to solve the problem of getting to 30 miles per hour.

We must want to capture our potential before we can have problems to work on. Most of the time, since we do not know what our potential is, we do not put attention on problem solving. If we do not really believe a full-time living is possible from a small farm we will not attempt to solve the particular problems to get there. I've actually had people castigate me publicly for suggesting that a full-time living from a small farm is possible. They accuse me of filling people with false hope. Can you imagine?

Only when we believe it is possible, that we are not performing at full potential, do we begin writing down problems, or hurdles, that we need to get over. Problem solving is limited by the goals we think are attainable. Most of us rock along in our little underutilized world because we don't dream big dreams. Nothing is a problem if we have low expectations.

How do we heighten our expectations? One of the best ways

is by brainstorming. The rule of brainstorming is that nothing is too wacky to go onto the list. The sheer brainpower that can be focused on a nonjudgmental list is incredible. Let's take out a sheet of paper and make a list of all the possible farming things we may want to do.

Just look at a few of the possibilities:

- cattle
- vegetables
- bramble fruits
- sheep
- milking sheep
- orchard
- flour
- pumpkin patch
- aquaculture
- nursery
- lumber
- nuts
- llamas
- rabbits
- chickens
- pigs
- turkeys
- flowers
- herbs
- firewood
- school tours
- jerky
- horse boarding
- goats
- mushrooms
- eggs
- organic soybeans
- grain
- garlic
- ostriches
- buffalo
- dairy
- fee hunting
- pheasant
- quail
- Christmas trees
- horse training
- veal
- crafts
- woodworking

I put this list down just as it came to me, off the top of my head, and it is as incomplete as your imagination. I'll bet you could add a bunch of things to that list. Or perhaps you would want to narrow down the descriptions a little bit. The possibilities are as wide as your imagination. Every time I think I've seen it all, sure enough some creative person comes up with a new angle, a new niche.

We can stimulate that creativity by making lists. We do this all the time, and then go back through to pass judgment on the options. That is how we finally arrive at realistic goals.

If we made another list of all the things you could do in the country to supplement your farming income, it would be too long for this book. It would include everything from mechanic work to interior design to home computer work.

Once we have a list, we identify specific items on which we'd like to concentrate. Often these will jump out at us, kind of an intuitive thing, welling up inside. This is where the hard part comes, because at this point we must submit those items to the scrutiny of judgment. We examine them with these elements in mind.

How quick is the outlay/income turnaround time? For any beginning enterprise, this is one of the most crucial questions. This is even more important than if it's our favorite thing to do. How many things have we had to do temporarily in order to accomplish a long-term goal?

That is normal. This idea that we're going to farm and everything is going to be roses and romantic is hogwash. Our most critical need is to pay the bills. That's the bottom line.

Probably the quickest cash turnaround time on the list is broilers: eight weeks from start to finish. Probably a close second is radishes or just any vegetable.

Even a quick fruit like strawberries is a year. If you will notice, most of the animals have a relatively quick turnaround time compared to other things. Some things that are not production oriented, like horse boarding or lumber, can be viewed without turnaround time except for the capital costs.

We need to fairly assess all capital expense involved and then look at a realistic payback schedule. Cash flow is the nemesis of most fledgling businesses, our farm included.

Do we have to borrow money to get into this? Borrowed money should make us think real hard about proceeding. It may be better to go ahead slowly with what we can afford than to go through the borrowing risk. Usually the amount we can spend will equal our experience level. In other words, if we need to borrow money to do it, we probably do not have enough experience to make it a risk worth taking. We call this principle "overrunning our headlights."

Once our experience level is high enough, chances are our purchasing power will be there as well. If we will let our investment ability dictate our decisions, we will not be tempted to spend beyond our ability to properly steward whatever we invested in. This includes buying land.

What is the market like? If the demand far exceeds the supply, go for it. But if the market is trending down, or if not too many people buy it, chances are the market is not good. I would include in this item finding out the financial status of current practitioners. Are these folks independently wealthy? Do they really depend on the farm for their income? These questions will penetrate the hoopla and get to whether or not the opportunity is good for us.

One way to find out the strength of the market is to listen to people's desires. "I wish I could find" is a key phrase. Whatever item comes at the end of it is probably in short supply. Another way is to think about what we want. What can I seem to never find but would love to have?

As you begin your farming enterprise, you will want to sit down routinely and make little brainstorm subsets about specific things you are doing. For example, let's say you are producing strawberries. You may want to make a list of all the things you could do to increase production:

- irrigation
- winter covering
- variety
- fertilizer
- foliar spray
- mulch
- more timely picking
- less slug damage
- less frost damage
- fewer weeds
- earlier fruiting

Maybe you need to make picking more efficient. You can:

- allow pick-your-own by reservations only
- plant different varieties to extend the picking season
- on low traffic days, give a discounted price to stimulate customers to buy greater volume
- on high traffic days, charge a premium to discourage too many customers
- limit the number of cars you will park
- offer prepicked strawberries at a big premium to stop some people from going to the patch
- offer good pickers a bartered 2-for-1 deal or something to stock the ready-picked basket
- change the layout of the patch
- offer entertainment that will divert some attention and allow you to handle more people at once — like a petting zoo or a small straw bale maze
- use above techniques to bring in more customers if that is the problem
- offer additional shopping opportunities by diversifying your market portfolio — eggs, other vegetables, meats, crafts
- offer children's activities like pony cart rides or supervised playground to reduce the number of children in the patch
- offer an alternative, cheap PYO to occupy children while parents pick berries — maybe some flowers
- use a different type of basket
- change your hours

Then you want to prioritize the most important to least important, including what will give the greatest benefit for the investment. Sometimes you can't do the one that will give the biggest kick because you just can't afford it. In that case, move to what is doable. We routinely establish objectives and then prioritize them according to time and money.

If you are not a list maker, force yourself to go through this exercise. You may be surprised at how powerful a tool it is. Writing it down rather than just thinking about it gets you to use additional perceptual senses, and that stimulates more brainpower.

One thing I have learned is that none of us is creative enough to keep from getting complacent in our own routine. Every time someone who is a good thinker visits the farm, I ask: "Okay, if this were yours, what would you do differently?" I have been amazed at the perceptions of other people, and have implemented many ideas based on their recommendations.

Sometimes the ideas are not good because they don't see the whole picture. But often they can see things I don't see. Being farther removed often gives a whole different perspective. When we're doing it every day, living in it, we often become satisfied rather than pushing to the next "big thing."

I would hate to be a conventional farmer and do the same thing every year. Here are some questions you might ask to keep from becoming self-satisfied:

Can I do this more easily without compromising results? A necessary component of this question is to stay on the point. If the point is functionality, then we're free to make whatever changes need to be made. But if our changes can only fit within the confines of accepted practice, we will never make the breakthrough improvement. Who cares what the machine or the tool looks like as long as it works? Eliot Coleman's functional garden hoe designs are a perfect example of this breaking with the norm.

One thing many people have encouraged us to do is to move the chicken pens with a tractor rather than by hand. Even if it were easier, it scares the chickens to death, so the results are not the same.

Can I do this more efficiently? If we're moving material, like carrying water, always carry two buckets rather than one. The cost is in making the trip. Go loaded and come loaded. If you find yourself driving or walking empty one way, what can you haul to make the trip more efficient? Part of this is trying to get multiple actions from a single action. For example, moving animals daily may take a few minutes, but look at the benefits: more forage production, better nutrient cycling, healthier stock.

Can I do this more simply? One of the curses of owning machinery is that once we have it, we become totally dependent on it. For example, I had a logger friend who needed to cut a couple of diseased trees out of his lawn. They were only about 10 inches in diameter and he was going to cut them up for firewood.

He complained to me that he couldn't do it because his skidder was up in the mountains on a logging project. I about fell over. The thought of using a skidder would never enter my mind. But in his mind, that is what you do to trees. You cut them down and then you hook up to them with this machine to pull them to where you cut up the logs. The thought of just backing a pickup up to the trees where they fell and bucking them up there with a chainsaw never occurred to him.

Can I get an animal to do this? Although this overlaps philosophically with the previous questions, it is fundamental enough that I think it deserves separate consideration. I think, generally speaking, that we have not scratched the surface on harnessing the innate characteristics of animals to make things easier for us. Here are some things that can be done with animals that are routinely performed by people or expensive machines (and usually both):

- pigs to till and turn compost rather than compost turners
- pigs to cultivate instead of mechanical tillage equipment
- chickens to spread out cow paddies instead of pasture harrows
- chickens to clean up dropped fruit in orchards, vineyards, and fruit systems
- dogs to herd
- sheep to mow orchards and vineyards
- turkeys to debug crop fields and forestal areas
- chickens to debug gardens

- geese to weed strawberries
- oxen to move anything
- staked or tightly occasionally confined milk cow or off-breeding season bull to mow tightly around outbuildings and lawn
- goats to mow brush instead of rotary brush mowers
- rabbits to mow the lawn
- chickens to slightly aerate garden soil as described in Andy Lee's book, *Chicken Tractor*
- compost and rabbits to heat a greenhouse
- cats to eliminate rodents instead of poison
- guard dogs to protect predator-prone small stock
- guard donkeys to protect sheep
- self-harvesting any crop rather than combines — hogs to eat corn and grain; cows to eat corn; poultry to eat small grain
- horses instead of automobiles
- pigs to till poultry house bedding rather than cultivators
- elephants to bring in logs

I put the last one in to see if you were still awake. Of course, elephants are used in many parts of the world for real agricultural work. I'm sure I've missed plenty of applications, but you get the idea. If we started mentioning draft power, we'd have another couple of pages. I don't know anyone who uses all of these, including us, but it's good to just see what's available. The theme here is to think *first* in terms of how an animal can do the job, rather than just taking the easy way out and using machinery or manmade concoctions.

People constantly ask me: "How do you think up all these things?" The answer is that I've been blessed to come from a long lineage of creative thinkers. My dad, and his dad before him, reveled in thinking outside the box. But beyond that, I have to answer that I am eclectic.

I try not to limit the sources of my information, and I try to visit as many farmers as possible. I've never set foot on a farm where I didn't learn something or notice some creative solution. Have you ever noticed how many creative gate latches there are? One of the most effective stimulants to brainstorming is just to expose yourself to as many ideas as possible.

That means subscribing to a wide array of periodicals, from biodynamics to permaculture to controlled grazing to marketing and the environment. Surround yourself with interesting people who will stimulate you in areas that are not within your realm of expertise. Cultivate friends who will ask you "why?" and make you articulate why you do what you do.

This will come naturally from your customers as you direct market. The more people you incorporate into your farm, the more angles you'll enjoy. Farmers generally limit themselves to ideas from people whose agenda is to sell them something, and that's a fairly imprisoning type of assessment.

Surround yourself with buyers instead of sellers, and a host of new ideas will come forth. Creative thinkers – brainstormers – are people people. Be one.

Self-Employment

One of the biggest barriers to folks leaving their employment and starting out on their own is the fear of losing all their work benefits. What a tragedy that our safety net now imprisons us.

Nonetheless, these concerns are real and have a great bearing on your ability to make ends meet. Let me share some fairly broad philosophical concepts to help you decide just how much of this safety net you want to carry over into your next life — the entrepreneurial farmer life.

Medical Insurance

This is the big one. The typical family now spends $3,500-$5,000 per year on medical insurance. Of course, the employer generally picks up a large percentage of it, which hides from people the reality of just how big this cost is.

Back in 1982, when I left outside employment, Teresa and I maintained a good policy carried over from my former job. But then things began to escalate dramatically. In just a couple of years its cost went from under $1,000 annually to more than $2,000. It was by far our largest expense.

We finally dropped insurance altogether. During that few years without any insurance, I had a couple of emergency room visits — remember I said I was a Type A? — and we paid the several hundred dollars cash for the stitches or whatever. We were concerned about not having any insurance, but also realized that medical care was not private enterprise anymore.

When I worked at the newspaper I covered "certificate of need" hearings. In fact, I covered a lot of hearings. I was always amazed at the number of regulations, licenses and government approvals required for seemingly unnecessary things. For example, if a hospital wants to add 20 beds, it can't just do it. Instead they must go through a costly analysis procedure to justify the requirement and describe what kind of care these beds would receive and on and on. Then a government agency would grant or deny a certificate of need for this expansion.

If I were unhappy with the current level of medical care in a locality and wanted to start a competing hospital facility, I couldn't just go out and find some investors, put together the money and build my idea of a good hospital. No way. I'd have to submit all this data showing how the current number of beds were inadequate, and show how my facility would meet current standards of care (peer approval) and all sorts of things.

In reality, what was once private enterprise is now a government entity. If the government were not involved with medical care, the prices would be a fraction of what they are because anybody could build a competitive facility. It works for fast food. It works for shirts and pants. But our society has decreed that we will not let market forces work in certain things like medical care, education and wildlife management.

What about quality of care? Probably some would have a poorer reputation than others would, but overall costs would be lower. And just as in bygone days, philanthropic organizations would help folks who couldn't help themselves. With the government out of the way, people would have more of their own money to spend or give away.

As a result of government involvement in medical care, I have come to the conclusion that demanding services without carrying insurance is moral. I certainly would not walk into a restaurant and demand food without being able to pay. I certainly would not walk into the tractor dealer and demand a gratis tractor. Actually, I can't think of anything I would morally feel comfortable demanding without paying, except for medical care.

The medical establishment has insulated itself from competition and has formed an alliance with insurance underwriters to abscond money from patients and has even conspired with doctors to eliminate alternative practitioners from being able to compete for the medical dollar. How many alternative doctors have had their offices smashed by government gumshoes?

In fact, the government now runs medical care to such an extent that the emergency room cannot legally deny me care if I walk in without a penny to my name. Normal businesses can deny merchandise to anyone, and that is the way it should be. But if a business insulates itself from competition and even colludes with government agents to run would-be competitors out of business, then it jolly well needs to submit to government requirements to run like a government agency, because in reality that is exactly what it is.

Once anyone can start up XYZ hospital, designed however suits the owners, dispensing whatever kind of elixir people want to buy, charging whatever prices patron and owner deem agreeable, then it will be time not to demand services gratis. But until that day comes, I refuse to feel sorry for hospitals that lament having to treat people who do not pay — you'll get no sympathy from me.

The bottom line here is that I do not think ill of someone who carries no insurance and yet demands services at the hospital. That tirade notwithstanding, something deep within my free market spirit hated to be in this position.

Then we heard about huge deductibles and it was like a breath of fresh air. We learned that as much as 70 percent of medical insurance premiums go for the first $1,000 in payments. Use statistics show that most of the premiums go for all these nickel and dime visits to the doctor.

Teresa and I dared to ask ourselves: "How would we use

insurance?" Two things came out of this question. First, we realized that our lifestyle astronomically reduced our chances of use. We had plenty of fresh air, sunshine and exercise. We grew all our own food. We even made our own bread. We didn't use alcohol or tobacco. Goodness, we didn't even drink coffee or soda. We home-school our kids so they don't have all the childhood sickdays that normal classroomed children do. That in itself is reason to home-school. From a healthy lifestyle standpoint, we lived an extremely low risk life.

Second, we realized that the only reason we would ever use insurance was for severe trauma. Let's say I cut an arm off with the chainsaw. Some major accident that leaves one of us with trauma was really the only reason we would use medical care.

We never went to the doctor and the kids didn't go to doctors. We used garlic oil for earaches and Vitamin C for colds. We also knew that if we ever got chronic ailments like arthritis, cancer or circulatory problems, we would not use the medical establishment. We would go to some other countries where they heal these things all the time. We would do chelation or whatever.

We have great connections for herbalists, naturopaths, homeopaths and alternative snake oil dispensers. We will not go the cut and burn route. Anything chronic will be treated with something that is either fairly cheap or would be considered quackery and therefore not covered by insurance anyway.

For anything except trauma, the conventional medical establishment, in our view, offered the deadliest advice you could get. It dawned on us that if all we would ever use it for anyway was severe trauma, our risk factor went way down. Our chances of ever using it were almost nil.

With that settled, we purchased a $5,000 deductible policy. A couple of years later we jumped that to $10,000 deductible when we accumulated enough liquid assets. That has kept our medical costs extremely low. In fact, our premiums have held steady because since we have not used it, the company annually increases the premium very little.

A couple of years ago I had a woodcutting accident and went

to the emergency room. Because I'd had a concussion, the hospital required that I have a CAT scan. Teresa and I looked at each other — we could just see the dollars mounting, but we weren't in a position to quibble. When we got the bill, it seemed awfully small. After doing some research, we discovered that the hospital has several billing tiers. When our insurance card went through their computer, and they realized we had a large deductible, they billed us at about 50 percent the stated rate because they were concerned about getting paid. Can you imagine?

Are you upset about medical costs in this country? Well, it looks to me like if everyone would go to a $10,000 deductible, we could cut medical costs in half. I know, I know, that doesn't hold water in the real world, but it still seems amazing that they would have that big a difference in the cost of a service. Can you imagine going into a Wal-Mart store and having the cashier drop the price of a pair of shoes by 50 percent because you had a wimpy credit card? I know that a large deductible like this may seem rash, but once you go through the thinking on it, I think you'll agree that it makes sense.

In addition, we have 400 farm customers who would gladly help us out if we had a problem. Isn't that what community is for? Our patrons give us things all the time. Not just little things, either. These folks would gladly step forward to help us out of a pinch if we vocalized a need. Our church fellowship group would also help us out in time of need, just like we would help out any of them.

The need for most medical care is lifestyle-related. If people would just quit consuming alcohol and tobacco we could greatly reduce that need. If we would quit eating as much sugar and stop eating fecal soup chicken and *E. coli* beef, quit using drugs and stop sexual deviancy or promiscuity, we would see huge changes in the use of medical care. When I see 50-year-old men with huge pot guts (we call it furniture disease – "my chest fell to my drawers") I have mixed emotions about their impending medical conditions.

I must admit that I fight the urge to say: "Look, when you have back problems and diabetes, it's your own fault." I know that's not a very loving thing to say, but I confess it runs through my mind.

Certainly all sorts of maladies and calamities occur to people who do not live risky lifestyles. But if those were all the needs requiring medical care, we would probably drop total medical expenditures in this country by 80 percent.

Some may think my position is foolhardy. I tend to think it is the lesser of many risks. You cannot escape risks in life. If you look at the statistics, hospital-induced illness makes a hospital one of the deadliest places to ever be. We must constantly weigh risks and decide which ones we will take. Driving down the road is risky, especially at 1 a.m. New Year's Day. We can minimize risk, but we can never eliminate it.

If we attempt to eliminate risk from our lives, we never have the joy of watching neighbors, church friends, social club friends and relatives receive the blessing of giving in our time of need. Maintaining the income level required to pay first-class insurance premiums is highly risky. How many people took the risk of a corporate job only to be downsized, outsourced and rerouted? I think we just need to understand that all of life is full of risk. Of course life is risky — you can die from it. We just need to decide which ones we're going to take and not worry about the others. Otherwise we will be paralyzed from ever stepping into the unknown.

The greatest heroes of history have always been incredible risk takers. I can't imagine Daniel Boone walking out into the wilderness, staying for months without any modern conveniences. What we don't appreciate is how much he knew about herbs, snares, tracking and a host of other things you and I don't know. That minimized his risk. I'll guarantee you that for me to head off into the woods like he did would be foolhardy. All the history books say he was not a good farmer, which was the most acceptable occupation of his day. He devoted himself to a different body of knowledge, was gifted in mapping, hunting and trapping, and actually earned a living.

Today, we can learn about a job, paying insurance premiums and pleasing the boss. Or we can devote our time and energy to building community, establishing friendship networks and self-healing techniques. The Amish have certainly survived quite well without medical insurance, but they devote their lives to different things than does the average modern American.

The bottom line on medical insurance, then, is that you will get treatment. You can reduce your risks to negligible levels by adjusting your lifestyle. You can devote yourself to a different type of safety net. Finally, a large deductible makes it affordable and is a compromise between protection and patronizing the system.

Automobile Insurance

Here again is another one of those big variables. The quickest way to chop car insurance in half is to eliminate the collision part. The only reason to carry collision is if the car is not paid for or if you do not have enough liquid assets to replace it.

Notice I did not say that you needed the money in your checking account to replace it. Certainly replacing it may pinch your budget for a couple of months, but that's better than paying an extra $100 per month on insurance.

You should never go into debt to buy a car. Period. Teresa and I have been married 17 years and have not yet spent $6,000 on automobiles, cumulative total, in that amount of time. We have a customer who is a successful real estate agent and he said the average couple has spent enough on automobiles in their first 10 years of marriage to almost pay cash for a home.

If you buy a $15,000 car and finance it for 36 months, you will end up paying $20,000 for that vehicle. If you have two, just look at the cost. When you figure that the average couple will trade those cars in 5 years, you begin to see the enormity of automobile costs.

I bought my first car for $50, after getting out of college. Teresa and I took our honeymoon in that car. I kept it two years and sold it for $25 more than I paid for it. How's that for depreciation?

We have never yet owned a car manufactured in the current decade. We came close this last time by buying a 1989 minivan in 1996, but the rule still holds. I hope we get a couple more years out of that minivan.

When you start farming, you will not be commuting to work

all the time. All you need is a $1,000 clunker and a neighbor who can work on it. When I returned to the farm full-time with that old car, we only drove 5,000 miles per year. With no more miles than that, a gas-guzzler is by far your most economical transportation. You will not put on enough miles to compensate in fuel savings for the higher sticker price of a newer vehicle.

Get the lowest amount of insurance you can get and drive an older car. The next policy is to have only one car. The two-car American family is almost sacred now, but it's not if you want to make a living on the farm.

If you can't trade trips to town with one vehicle when neither of you is working there, then you must be way too busy. I hate going to town. I'd be perfectly happy to stay right here forever. Teresa would be happier too, because every time I go I spend too much money. I let things stack up and stack up until when I finally do go I purchase lots of things. Teresa says she can only afford to let me go to town once a month.

We've always gotten along fine with one vehicle. If it is out of commission for some reason, we borrow one from a friend. That's what community is for. If you don't have any friends, get some by being friendly. Anyone who can't get access to a vehicle in a pinch is a hermit indeed.

Take good care of the car by being religious about lubrication — changing oil, grease, transmission fluid, etc. If you currently own a couple high dollar vehicles, sell them and buy a cheaper one. Not until you are willing to do that are you willing to even think about a profitable farming enterprise.

All of these things will reduce your auto insurance — fewer miles driven, older model, no collision, and only one.

Homeowners' and Farm Insurance

Here again, the insurance premium reflects the value of the assets covered. Thousands of people who decide they want to farm buy some land and then build a $75,000 house — or even bigger, in many cases.

Do you want to put on a show, or do you want to make a

profit from farming? If you buy land, move on a used mobile home for $7,000. That will cut hundreds of dollars off your insurance premium. Becoming a member of the landed gentry takes a pile of money if you're going to do it fast. And if you do build a fancy house first, don't complain that there's no money in farming. For a summer you can live in a hut.

Downsize to what you can live with, and then put all your equity into capitalizing the income-producing aspects of the farm. That's really what it boils down to — deferred gratification. I'm not saying you should never have a fancy house, although I can't think of a reason why anyone should. But if you want to farm you must be willing to keep your focus on what takes money and what makes money. A house does not make money. A car does not make money. Once they are bought, you can't unbuy them. Dollars you never spend are still up for grabs; once they are spent, you're locked in.

Farm buildings are the same. Crude, functional, low-cost buildings need not be covered like buildings erected by the local construction company. Obviously if you have very little equipment, your insurance will be commensurately less.

Shop around for coverage. You would be surprised at how big a difference exists especially among farm underwriters. Find an agent who wants to become a customer, and if all the figures look acceptable, go with that agent.

If you would not replace the structures with planed lumber and hired labor you need not insure the buildings against that cost. Insure only to the equivalent replacement value. Again, remember what you are insuring against — catastrophe. The single biggest reason farm buildings burn is because rodents gnaw electrical wiring. That is why years ago we cut off the power to our barn. That just eliminated fire unless it's struck by lightning. If we do have a catastrophe like a tornado, our church fellowship community and our neighbors will come and help us rebuild.

We home-church, and our fellowship actually schedules workdays at each other's places. We've built barns, cut firewood, cleaned cisterns and demolished buildings. That, my friend, is better insurance than any money can buy. By not maintaining an institutional church building, we are freer to help each other rather than devoting time and money to maintaining the building.

All of this may sound hokey to sophisticated self-made urbanites who tool around in sport utility vehicles to art shows, soccer games, and the computer cubicle at work. But for those of us who relish a different life's direction, a lower-cash, community-based, simpler life, it is not a sacrifice, but a reward.

Taxes

I am not an accountant, but we use an accountant — an aggressive one. He is not just an accountant. He is a friend and promoter. That is important. He really understands what we are trying to do and looks for ways to help.

I'll never forget when we had a run-in with the meat and poultry inspection folks. The officials set up a meeting with us at our house and we called our delegate, senator and attorney. Our delegate sent his legislative aid, the senator sent a letter, and the attorney came to the confab. When it came time for introductions, our attorney, in his smooth baritone voice said: "I'm a friend, a patron, and an attorney." Having expert opinion from folks who are more than invoice-mailers is special — cultivate it.

I can't begin to get into all the tax laws — if I did, they would probably be outdated by the time we can get this book printed. Be assured that accountants run the gamut of aggressive and timid. Everyone knows that the tax code has lots of gray areas. You can ask ten IRS agents about an item and get ten different answers.

Any tax code as voluminous as the one we now have is obscene. But that's what we must live with. Find an accountant who really wants to save you money, not one that wants to make sure the IRS gets every possible penny. Yes, you need to pay what is due, but that is all.

If you are into direct marketing, as an entrepreneur, you can legitimately take far more liberties with expense deductions than just a typical wholesale commodity farmer. Be sure to keep receipts on everything. You may want to visit a financial planner to find out about incorporation or other shelters that can help you keep more of your money.

I do not believe that staying under the poverty level for your whole life to keep from paying any taxes is a worthy goal. You should relish the fortune of joining the privileged one percent of full-time farmers who actually pay taxes. That is an elite group.

But remember that if you grow all your own food, all your own fuel, and have almost no medical insurance or automobile costs, you can live extremely well on a poverty income. Just because your tax return puts your income at $10,000 doesn't mean you can't live equivalent to a $40,000 urbanite. Appreciate these cashless perks. In the early days of your farming enterprise, you will need all of them you can get.

I know people who doggedly prepare and file their own taxes, and I appreciate the savvy to do that. But I also know how much a professional preparer can help, not only in structuring your paperwork but also in knowing about special just-enacted exemptions or credits. I've chosen not to bog my life down with those things, and once things get more complex, my vote is you'll be money ahead to do likewise. You need all the creative energy possible to think up a better tomato trellis or a better marketing strategy. You don't need to drain off those valuable energies keeping up with the latest tax regulations.

Schedules

Although this may not seem to fit in this chapter, I think it does because it is one of the biggest unforeseen problems facing self-employed people. There is a saying that a self-employed person is one who works 80 hours a week for himself in order to keep from working 40 hours a week for someone else.

While that may not be totally true, it certainly is more true than not. Self-employed people have to hustle and answer to themselves. Many of us need the prodding of someone else to accomplish things. When the boss gives us a deadline, or the time clock needs punching, even undisciplined people can rise to the minimal expectations. But when none of those outside prods exist, we can easily fall into undisciplined habits.

We have a wonderful customer who was CEO of a Fortune 500 company. He is now 95 and although his gait is slow and he wears a hearing aid, his mind is as quick as ever. Even though he doesn't need to be anywhere, he still gets up early each morning and puts on a bow tie and suitcoat. Dressed just like he always did to go to the office, he enters his study and does his deskwork in his home. Here is a disciplined man. He could wear sweat pants and moccasins. With his money, he could sleep until noon every day. But here he is, punctual to a clock tick, dressed to the hilt, conversant on all the latest business and political details. Here is a man who challenges me; what a privilege to know him.

Too many folks have the impression that anyone who doesn't work for someone else is retired. I've been asked to do things in the community because "you're free." Sure, sure. I feel like responding: "You just come and walk in my shoes one day. Let's see how long you last, buddy."

The notion that people like us just glide around from effortless task to enjoyable entertainment is a common misconception. When you work for yourself, you have to stay ahead of the pack either by thinking smarter, working harder, or both. That requires self-discipline and scheduling.

The romantic view that you can leave the city, go to the country, have breakfast on the veranda at 9 a.m., then mosey out to the chicken house at 10 is devastating to a would-be farmer. By that time, half the chickens will have laid their eggs — on the floor because you didn't open the nest box.

Gardeners know that in the hot summer the most productive working time is from before dawn until about two hours after dawn. When we make hay, we try to do all our unloading in the haymow from the first streaks of daylight until dawn.

Getting into the rhythm of natural cycles, watching the weather to know what jobs to do, being timely about weeding — these can make or break a new farmer. It's not okay to milk the cow at 4 p.m. one day, 5 p.m. the next and 3 p.m. the next. It's not okay to leave the glass down on the cold frame when the temperature rises to 110 degrees. It's not okay to leave the peppers uncovered when the weather report calls for a frost. It's not okay to delay pruning the grapevines until warm weather stimulates heavy bleeding from the cuts.

Keeping a schedule and efficiently staying up with what needs to be done day to day is perhaps the hardest thing for a new farmer to learn. If you don't get those beans weeded today, what will you do tomorrow when the squash needs mulched? Better forego that second glass of tea at lunch — or forget lunch altogether. Sometimes you will sprint and sometimes you will relax. When Dad sent us boys running to the shop for a wrench, he would always say: "Run now; you can rest when you get back." We have periods when we work hard, but we also have long periods, like winter, when we don't do much. One is the prerequisite to the other and we must be careful to not confuse the two.

I have a little routine I go through with each new apprentice to help bring this home. I'll walk with him out through the field, perhaps going to the chicken pens or out to the cows. Ninety percent of the apprentices, or visitors, will lag behind, perhaps 2 yards. We'll walk maybe 200 yards and they stay 2 yards behind. What that shows me is that they are capable of walking alongside me, but they simply don't. Have you ever tried to converse with someone walking 2 yards behind you? It slows you down, you have to turn your head — it's dreadfully awkward; so much so in fact that you tend to just shut up after the fifth: "Huh?"

Needless to say, one of the first principles every apprentice learns around here is this: "When Joel walks, you stay right with him. If you have to run, so be it." A farming enterprise is not something you just lolly-gag around. Make a job list at the beginning of the day to make sure everyone stays on task.

Whenever we feel a little overwhelmed, we all sit down and

take 15 minutes to make a job list. It's the best investment we ever make. The rest of the day runs smoothly and we look back at the end of the day amazed at how much got done. Some days kind of take care of themselves if we're on a huge project that requires all hands all day. Teresa is a better list maker than I am. She's so good that if she does something that isn't on the list, she adds it to the list so she can mark it off. Now that's what you call task oriented!

When folks decide to move to the country, it often does not involve earning a living from the farm. That's fine. But recognize that their lifestyle will not be yours. People who retire to the farm and spend their leisurely morning drinking coffee, watching *Good Morning America* and reading the paper are not the role models for poor, hungry, entrepreneurs wanting to create wealth. These retired folks have put in their time; they earned their privileges.

Dad was a wonderful role model in this whole area. He had lots of interests and easily got sidetracked reading all sorts of things. When he felt like he was spinning his wheels with the farm work, he would record, down to the minute, how much time he spent each day doing different things.

The first time I remember him doing this, I was in high school and he was still working in town. He would come home from the office and sit down to read the newspaper and whatever else came that day. I'll never forget all his sheets noting the times he started and stopped things — even brushing his teeth. After a week, he made a grand announcement at the dinner table: "From now on, I will not read when I get home from work until *after* dark. I spent 12 hours last week reading the paper. That's a whole day of work lost during daylight hours."

I remember him doing it a couple times later and it was always quite a revelation. If you feel a little weak in this area, I would suggest such an exercise just to self-discover where your time goes. This may seem elementary, but if we don't get these fundamentals down right, we can never refine to better things. Life is built on doing the most basic things first.

As you begin your farming enterprise, the most basic thing is to become fully employed in income-generating tasks. This is not as arduous as it sounds. When you see the farm making money, you will enjoy it a lot more.

What are the big differences between working for yourself and working for someone else? Taxes, insurance and scheduling responsibility. Only as you successfully transition in these areas will you enjoy the emotional, economic and family benefits of self-employment. Don't be discouraged; be challenged. The change and growth process will develop you into the person you always wanted to be. It's worth it.

Evolving Your Model

Grass is the Center

Nothing builds soil like perennials. Nothing destroys the soil like tillage. On any farm, we should maximize soil-building principles and reduce soil-destroying principles. That means grass and trees need to be encouraged. Even grain farms should focus on this principle.

You may ask: "Who is going to grow all the grain, then?" My answer is simple: "Seventy percent of the grain we grow goes through multi-stomached animals that God made to *not* eat grain. We should actually reduce our grain production by 70 percent."

"But I thought grain was the foundation of an agricultural system," you say.

Let's look at that a little bit. Every animal eats grass, but some eat more than others do. Here are some typical farm animals and the relative percentage of their diet that can be acquired from perennial forages:

Cows	100
Sheep	100
Horses	95
Rabbits	65
Turkeys	50
Pigs	50
Chickens	20

If you will look at modern American agriculture, however, you will notice that this is not the way we feed these animals. For the most part, grass fed to animals is mechanically sheared, packaged, processed and stored before being fed. By and large livestock and poultry do not receive any grass, and certainly do not receive any direct-harvested, fresh on-the-stem grass. Although cattle and sheep do graze, this accounts for a relatively small part of their diet. From grain supplementation to silage to hay, even these ruminants do not receive 50 percent of their diet from self-harvested grass.

Is a grass-based approach really important? Yes, and here are some reasons why:

Nutrition. Fresh greens contain tremendous amounts of B vitamins (anti-stress and nerves) and natural antibiotics. These are not only important to maintain health, but are also extremely volatile. As soon as grass is cut, large amounts of these components vaporize. That is why feeding hay or green chop is never as nutritious as allowing the animal to direct-harvest the forage right off the stem. Wilted forage is not nearly as nutritious as the fresh material.

Vitamin-mineral to calorie ratio. A high calorie ration can certainly make things grow fast, but often the growth is not healthy. Imagine going on a Snickers bar and ice cream diet. You might grow real fast, but you would not be healthy. Even the fact that you might enjoy it for awhile would not make it good for you. As we increase the energy and reduce the vitamin-mineral portion, our high performance is masked by increased veterinarian bills, fertility problems and unthriftiness. This, of course, is exactly what we're seeing in confinement factory farming. The cost of trying to stay ahead of the latest disease problem or physical malady threatens to undo some of the largest corporations.

Exercise. Any animal allowed to graze fresh forage will automatically get exercise, a necessary element in growing healthy joints and muscle. When animals are penned up all day either in crowded

Photo 22-1. *Son Daniel moves a "Harepen" to the next spot on pasture. The grass reduces feed costs on weanlings, amounting to a value of $4,500 per acre per year. All animals should receive as much perennial salad bar food as they want in order to have healthier stock, more nutritious meat and reduced supplement costs.*

Photo 22-2. *Daniel lifts a "Harepen" and shoves a set of metal skids, or shoes, underneath to allow him to move the pen to a fresh salad bar.*

conditions or in tight cages, they do not move around. Movement keeps digestion going, keeps the blood flowing and reduces stress.

Fresh air and sunshine. Out on the grass, animals can breathe clean air and soak up sunshine, two critical elements denied 90 percent of the farm animals in this country. In a feedlot or factory house, animals can only breathe air laden with fecal particulate and dust. A pall actually hangs over feedlots. In confinement poultry and swine houses the ammoniated air is too toxic to breathe. Sunshine contains full spectrum light, which is essential to stimulate certain glands. Without full spectrum light, certain glandular functions become lethargic and this inhibits nutrient metabolism.

Soil building. The principle of "new ground," found in any agronomy book, is recognition by soil scientists that any crop following a rotation of sod will perform better than one following a tilled crop. Old agricultural books typically encourage multi-year rotations in which hay occupies the premier spot. Hay, of course, is dried grass. A common rotation is corn-oats-hay-hay-hay-corn. Variations on this theme abound, but years ago anyone with common sense recognized the premier position of grass in a cropping or livestock rotation.

Before 1950 any livestock husbandry book had pictures of portable chicken houses on skids, pastured hogs, pasture farrowing huts and dairy cows grazing white clover/bluegrass pastures. Beeves fattening on grain also had access to pasture. The grain finishing was only a supplement to the pasture.

Nothing replaces pasture for soil building capability. It stimulates earthworms by allowing their burrows to remain intact, rather than be destroyed through tillage. It provides a haven for moles, voles and field mice that feed predators and aerate the soil. It maintains a ground cover, a protective vegetative blanket, over the soil to reduce erosion and actually build more soil than is lost. The mulching effect stimulates biological activity by protecting soil life from the harsh sun and by holding in moisture. Little soil microbes actually slog around in a microscopic marsh of water, minerals, and vegetation.

With all this in mind, when I hear someone question how we can possibly convert a large percentage of our farm to grass and stay in business, my response is I don't see how we can afford not to. Obviously the requirement is easier to see if we have livestock, but it is also true for any cultivated crop type operation.

While I do realize that some organic vegetable and grain farmers have been able to maintain fertility without a grass rotation, I also know that it takes a huge amount of diesel fuel and other inputs in order to do it successfully. In my view, we really need to be reducing our crop acreage — even as much as 70 percent. This will open up huge portions of land for the kind of long-term grass healing I'm talking about.

Economics. Now we get into the meat of this discussion. We have a grain-based livestock production paradigm in this country — indeed, in all developed countries. One of the reasons I am keying in on livestock here is because animals consume so much more of the production from crop ground than do humans. If we could fundamentally change our livestock systems, the tilled acreage requirements to produce people food would diminish to the point that blocks of tilled acreage would be small enough to be at least benign on the landscape. The surplus soil-building capacity from the rest of the acreage could then be devoted, in a non-petroleum-based, sustainable way, to these few acres to keep them in a high state of productivity.

I am certainly not opposed to tillage for vegetables or grain — I don't purport to have non-tillage answers to these crops. We can, I am convinced, do much more with mulches and smother crops to reduce tillage, but I do not know of long-term models right now that would allow us to produce crops over the long haul without some sort of tillage. I am simply examining this issue because the crux of this book is to make *you* successful.

We need not deal with all the maddening questions like "What would we do with all the poultry houses?" or "What would happen to John Deere?" or "What would happen to the value of all that center pivot irrigation equipment?" The focus here is on *your* success.

Obviously if our agriculture went through some fundamental systemic changes, the new criteria for success would change. The goal here is to examine what the profitable alternatives are in the current paradigm and how you can fit in the picture.

With all that in mind, let's look at the economics of a typical 100-acre Midwestern grain farm compared to one producing grass and putting that grass through beef steers. Let's assume the land is fertile, perhaps 150-bushel per acre dryland corn. If corn is selling for $3 per bushel, that's a total gross income of $450 per acre. Now we need to subtract seed, fertilizer and fuel: about $150. That leaves us $300. Now we need to take off either the cost of tillage, planting and harvesting by a custom operator, or the equivalent equipment costs if we own a tractor, plow, disk, and corn planter. Let's assume we hire a combine to come in and harvest the corn. Now we subtract another $150 per acre, and that leaves us with $150 out of which we need to pay property taxes, insurance on buildings, mechanic labor, parts, and the like. Now we're down to less than $100 per acre to actually buy groceries, fix the house, pay life insurance, keep a car, clothe the children and live. Obviously, we're going to have to take a job in town in order to make ends meet.

Now let's take that same farm and put it in grass. It has no buildings. It will produce 600 cow-day grass (a cow-day is what one cow equivalent will eat in one day—this changes depending on season, lactation, etc.). That amount of grass will feed, during the spring, summer and fall — let's say, 7 months — approximately 200 stocker calves (210 days at 600-cow-day grass on 100 acres). We'll buy 500-pound calves for 80 cents a pound, which is a total cost of $400 per calf, or $80,000 total. We'll assume an average daily gain of 1.3 pounds, which means we'll put 260 pounds on those calves and sell them for 75 cents a pound at 760 pounds, or about $570 per calf, for a total of $114,000. After we subtract our initial purchase price of $80,000 we're left with $34,000. We may need to pay some interest if we borrowed the money, and we probably used $2,000 worth of fence and water line. We used not one piece of machinery because this is such a small farm we can walk anywhere on it.

So if we take off $4,000 for expenses (fence, water line, minerals, vitamin supplements, fuel), we're left with $30,000 to fix the house, buy a car and take our family out to dinner. In *Salad Bar Beef* I detail the nuts and bolts of this production system, but suffice it to say here that there is simply no comparison between typical grain farming and intensively managed grass farming. Now let's ask the obvious question: "What happens to grain prices when lots and lots of people see this opportunity and convert to grass?"

The answer, according to Allan Nation, editor of *The Stockman Grass Farmer,* is this: "Grain will become extremely expensive because in order to get people to grow it, they will need two or three times what they get now. At that point, you will see poultry more expensive than beef. Once the true costs of grain production are taken into account, grass-fed beef will become the new environmentally-friendly, low-cost protein source."

Obviously every 100-acre farm will not produce this kind of return. One in the middle of the desert will not do this; of course, it doesn't cost $4,000 per acre, either. But the basic comparison on economic and production potential holds true in all regions. And sometimes the spread between purchase price and sale price will be greater, which would reduce the bottom line. Sometimes it is closer, but that would normally mean neither is high. The calves may gain better, too, which would increase the bottom line. This sample is certainly subject to variables, but the basic pitch is correct.

The farming opportunity in pasture-based livestock enterprises is incredible because typically farmers do not put any attention on grass. They talk about "the back 40" as if it is a scourge. And, of course, that's where the pasture is — abused, neglected, unproductive. Literally thousands of idle acres, in various stages of disuse, dot the countryside and await the stewardship of someone who knows how to take advantage of the magical union between herbivore and grass.

Grass does not need to be replanted. It can be harvested with animals that are appreciating instead of machinery that depreciates. It need not be stored, handled or packaged. It is not prone to insect or disease damage and therefore is perhaps the lowest risk of all crops.

Food quality. The taste, texture and nutritional profile of any animal protein — poultry, dairy, meat — produced on pasture is off the typical USDA charts. The superior quality assures you a premium price and customer loyalty. Meats and poultry have less saturated fat; eggs actually reduce cholesterol; meat has higher vitamin content. The list could go on and on, but the bottom line is that the differences between factory-based and grass-based animal proteins are measurable and obvious by a wide array of standards.

Politically. Grass-based systems are clean and green, buzzwords of the new millenium. Our communities want clean industry, clean employment. Folks want the same in their food and livestock systems. The difference between seeing a bunch of chickens out on clean pasture compared to seeing them locked up in a confinement house is the difference between having something you'd like people to come and see and having something you hope nobody will find out about. The aesthetics and the emotional perceptions of the two models could not be more opposite.

Suppose we do want to plant grain or vegetables. What then? Incorporating grass into large percentages of the acreage can save money in fertilizer and bug problems by creating a healthier soil. The grass acreage need not be seen as an unproductive use — even for a vegetable farmer. Our broilers net $1,200-$1,500 per acre, which may not be equivalent to vegetables, but it's not shabby either. I detail this model in *Pastured Poultry Profits: Net $25,000 in 6 Months on 20 Acres*. If we intensify the livestock component on these grass acreages, they can remain profitable.

Usually crop farmers assume that any grass acreage is doomed to low productivity. It is basically wasted land. But with low-cost, intensive livestock management, this is not the case at all.

We must view ourselves as earth stewards, not cattle farmers, grain farmers or vegetable producers. The whole mentality of viewing ourselves as species-specific practitioners destroys the holistic mindset that sees the farm and, indeed, or whole lives, as being a non-compartmentalized whole. We must stop this mental divorce

Photo 22-3. *Polyface apprentice Joel Bauman from New York moves a pastured broiler pen to a new grass salad bar. The chickens walk on the ground to the fresh greens. The daily moves maintain a high level of grass ingestion because the forage is not soiled from the previous day. This produces a superior product with designer qualities.*

Photo 22-4. *A specially-designed dolly, acting as a portable axle, allows pastured poultry pens to be moved efficiently to fresh grass.*

and appreciate that as farmers or would-be farmers we are caretakers first and foremost. Within that context we will find things at which we excel, and things on which we may want to concentrate. It would be similar to a music major not being required to know anything about writing or speaking. In colleges, music majors always have a proficiency. "I'm a music major with proficiency in flute." That means the student takes history, language, some math and science, along with music courses, but has a proficiency, or a special concentration, on playing the flute. Just because you have a proficiency to grow vegetables does not mean you know nothing about livestock or pasture management. Nature is diverse, and it is a whole. Let's be done with compartmentalized thinking.

Using hogs to incorporate mature cover crops, for example, is a wonderful wedding of vegetables, tillage and animals.

The earth's grasslands, building literally feet of soil over many centuries, need to work their wonder on your farm and mine. Putting grass on a pedestal, and in fact using it as a focal point of our agricultural enterprise, can help ensure long-term productivity, profitability and distinctiveness.

Biodiversity

Perhaps the most fundamental aspect of landscape design is biodiversity. Whether you have a window box, a backyard or a million-acre ranch, the more diversity the better. Diversity is the balancer, the stabilizer, of any landscape. Fire, flood, wind and other natural catastrophes maintain diversity by harvesting, covering and changing. Diversifying the landscape builds in stability and balance.

Any observer of natural ecosystems realizes the redundancy that is built into the landscape. To keep the soil covered, vegetation grows from seeds carried on the wind, in bird droppings, on fur, hair, feathers, in water and on insects. If one doesn't make it, the others will.

The three great environment types are open land, forest and water. Certainly these meld together in marsh, savanna and mountain treelines, but in general, those are the three basic environments. Most species require two of these to thrive; very few thrive in only one. The more we can intersect these three environments the more diversified both plant and animal life will be. Game biologists call this phenomenon "edge effect." The more edge effect we can create, the more varied species proliferate.

The industrial mindset destroys diversity and pushes everything toward standardization. Bulldozing out the woods in order to

Photo 23-1. *Edge effect should be encouraged by fencing out wooded areas. This offers diverse habitat for miles along pastures.*

make perfectly square cornfields, for example, mocks natural undulations of the land. It also destroys natural balances in the agriculture sector.

For example, birds, which are God's pesticides, generally do not feel comfortable feeding in areas farther than 200 yards from cover. They will fly over an area farther away than that, but will not settle down and feed. They feel vulnerable too far from home and hearth just the way you do — unless you're staying with friends.

When miles and miles of cornfields, unbroken by woodsy or brushy cover, dominate the landscape, birds have no place to live. They cannot come from 50 miles away to eat the moths and corn pests. Nothing is as sterile as a cornfield, especially an herbicided, weed-free one, especially if it is in the middle of a hundred other cornfields just like it. Pioneer accounts of birds indicate that our flying friends have been decimated over the decades. First, farmers try to destroy their food supply by eradicating all the bugs with chemical sprays. Then farmers proceed to destroy all the birds' nesting places by creating huge monoculture fields rather than small fields broken up by little patches of woods, shrubs and water.

At our farm, over the years, we have created forestal runways into our open land in order to maintain this 200-yard rule. In our area, forest zones can be created both by planting trees and just abandoning an area.

This brings us to the discussion of succession, probably the most powerful natural force. Life is dynamic, never static. If you quit mowing the lawn, certain changes would occur. The grass may grow long, then some shrubs would appear, then trees. That would be a natural succession in temperate parts of the country.

In other areas, like dry climates, the grass would first grow tall, then implode on itself and eventually die out. Everything is in a state of flux. A pond you build today will eventually have aquatic plants around the water's edge. In a few decades it probably will fill with silt and turn into more of a marshy area. A mature forest eventually builds up enough dead and dying material that it is susceptible to wildfire, just like the prairie. This is nature's way of harvesting and rejuvenating.

People who think they can take an ecosystem and freeze it for years to come are not being biologically honest. We as thinking people can either push succession backward into a debilitating tailspin or we can push succession forward, actually stimulating a healthier landscape through our ingenuity. The work of P. A. Yeoman's keyline system is an excellent example of this. His book, *Water for Every Farm,* outlines his landscape-enhancing techniques.

This natural propensity toward succession and diversity affects everything we do. A pasture reverting to thistles and shrubs is in a state of negative succession. The trick for the grazier is to make succession work positively through nutrient cycling, grazing management and energy flow in order to maintain highly productive and palatable pasture species in the sward. I deal extensively with this issue in *Salad Bar Beef.*

Weeds covering bare ground spots in the garden signify nature trying to clothe the soil. Bare soil is extremely unnatural. That is why tillage is one of the most devastating things we can do, and why it takes such energy to maintain tilled ground in a high state of

Photo 23-2. *Running fences on the keylines capitalizes on topographical diversity. At Polyface, we have no straight fences. Letting the "lay of the land" determine use and boundary makes fantastic landscaping for both form and function.*

Photo 23-3. *Keylines, roughly defined, are the break points between slope and swale or slope and ridge. Here a fence runs along a swale.*

productivity. Unclothing, or devegetating the earth's skin opens the soil to sun, wind and rain drop bombs. The less we denude the soil, the easier it is to maintain fertility.

Tillage, then, should be done in patchworks of small areas buffered by protective ground cover. Large, tilled, monocropped areas spell environmental disaster, and require huge energy inputs to remain productive.

To stimulate diversity, and its corollary succession, think about how you can break up the landscape with productive forms of the three great environments: forest, open land, and water. Obviously the water garden craze is a step in the right direction for lawns. But more than that, lawns should be turned into food production places. Think of all the land going to waste in this country around rural residential estates just to provide a place for people to drive lawn mower tractors around. If you happen to own one of these places and have a hankering to farm, I suggest you turn it into a diversified farmette paradise.

When I think of what my grandfather did on scarcely half an acre, I refuse to accept the idea that the first thing you need to do in order to farm is to buy a big plot of land. Plant vineyard, orchard, vegetables, bramble fruits, put in a rabbitry for a good source of manure and get a backhoe in there to dig a pond. If it won't hold water, line it with plastic. See how many species of plants and animals you can put on that little piece of ground and then, if and when you do outgrow it, you will be ready to be entrusted with a larger acreage.

Consistently and systematically harvesting wooded areas is one of the best ways to ensure biodiversity in those areas. Trees grow old and die. They are renewable things. When you cut one down, a new one can take its place. Climax species and pioneer species have distinct characteristics and grow under widely divergent conditions. The same is true for deciduous and coniferous species. It is important to be aware of a few broad silvicultural principles:

Light. Perhaps the most important element in forest management is how much light penetrates the tree canopy and strikes the forest floor. Different species are more or less shade-tolerant. Deciduous trees with big leaves, like oaks, hickory and gum, are extremely shade-tolerant. They thrive on less than 50 percent sunlight.

Conifers are at the other end of the spectrum, and are shade intolerant, meaning that they need a lot of light. That is why conifers sprout following violent forest fires that kill the hardwoods. Conifers are also naturally concentrated on south slopes rather than north slopes, especially in the mid-Atlantic region of the U.S. The southern slopes offer more sunlight than northern slopes.

Deciduous tree trunks are covered with dormant buds that can sprout if exposed to too much light. Sprouting is called epicormic branching, and destroys the value of the wood. This is also called suckering, or some foresters will talk about a tree getting "fuzzy." This all denotes little branches coming on an otherwise limbless, straight growing deciduous tree.

The more light that strikes the forest floor, the more brush and shrubs will proliferate. Brambles require almost full light, and often encroach on freshly harvested acres. From a diversity standpoint, therefore, one of the best ways to have some native bramble fruits is to open up wooded areas by making small clearcuts. This will automatically change species composition for a few years until trees again get big enough to shut off the light to the forest floor. At that point, the brambles and other light-sensitive, pioneer species will die.

Site index. This term expresses the growth potential on a particular site and usually applies to certain desired species. The higher the site index, the more basal area an acre can grow. Basal area is the area, in square footage, occupied by a tree when measuring diameter at breast height (DBH). A given acre of land can only support a certain amount of basal area. It cannot be completely covered with trees.

If you have a site index of 60 and a basal area of 90 square feet, the greatest productivity will occur if you thin down to 60 or maybe to 50, and try to keep the basal area in the 60 range. The problem is that in many areas, especially the mid-Atlantic region and perhaps the good hardwood regions of the upper mid-west, forests have been high graded several times. That means the best trees have been cut and poorest left.

Obviously when that happens several times, the quality of the trees decreases because the prime ones are consistently harvested and the poor ones are left. Many sites, therefore, are incapable of adequate thinning because not enough good trees remain after thinning out the junk ones to control light. If you thin too heavily, remember, the remaining trees will sucker and shrubs will proliferate.

Furthermore, harvesting just a few trees is difficult without skinning some of the residual stand. Once a tree is wounded, it never recovers fully and will often die in a few years. When it dies and falls over, it may well bend over and deform several nice saplings trying to grow to fill the void. More often than not, I favor small clearcuts to simply start over completely. This encourages maximum diversity.

About ten years ago I tried thinning a nice oak site. Many of the nice stems I left started sprouting. As the hill rolled over to a more southern aspect, there simply was not enough good stuff to leave, so I simply cut the entire area clean — not quite half an acre. Now, ten years later, we're going back in and clearing that thinning to salvage what we can before the remaining trees sprout any more. The little half-acre spot, however, has an aspen grove (very unusual in this part of the country) and has grown gallons and gallons of blackberries and blueberries. Now nice oak saplings are up 12-15 feet tall and beginning to shade out the berries in the understory. It is more beautiful, more diverse, and far more productive than the area we tried to thin.

At another 2-acre site we noticed a fair amount of white pine and gnarly old oaks. The southern slope was too hot and the soil too rocky and dry to grow good hardwoods, even though some had managed to eke out an existence. In that area, we cut out the hardwoods and left the white pine, which have since jumped due to the full light. They are well suited to a hotter, drier site and will be healthier than the oaks ever were.

About 15 years ago we had a Virginia pine site that was an abandoned row crop field from 50 years ago. It is a steep northern slope and was finally abandoned when gullies made tillage impossible. Today it has gullies anywhere from 4-10 feet deep running like corrugated roofing down the slope. The pine trees, which are pioneer species and only live for about 30-50 years before they begin leaking pitch and die, were beginning to decline. We cleaned off the half-acre area to stimulate the production of hardwoods that had begun to sprout in the shelter and shade of the pine overstory. After we cleaned off the site, these suppressed locust, oak, poplar, ash and walnut saplings took off. Within one year these little trees grew an average of 8 feet. Of course, for the first couple of years we couldn't even walk through there because it was so full of brambles and vines. Deer loved it. Today it is a beautiful stand of 4 to 6-inch hardwoods, easy to walk through and incredibly productive.

About 20 years ago I cut over a 3-acre bottomland wooded area, basically doing a light thinning. Brush did proliferate but because of a steep northern bank that shaded the area, the trees did not sprout. That area today has some fine poplar that we can now harvest for sawtimber.

In our area it is common for farmers to purchase pressure-treated posts for fence building. In years gone by, no farmer would have dreamed of such a thing. With systematic woodlot harvesting, a constant source of locust trees supplied all the poles and posts a farmer needed. The fact that farmers are now buying these things only demonstrates the lack of diversity being maintained in the woodlots. Locust trees are pioneer species that have a short life. They only grow in open areas where they can receive full sun.

Because we are constantly cutting little patches of woods, we have plenty of locust trees coming on for the next generation. In fact, we have enough that we can sell some to the neighbors.

Board feet. Lumber is measured in terms of a board foot, which is a block of wood 12 inches by 12 inches by 1 inch. Plenty of tables and measuring sticks exist in the forest industry to provide a quick ballpark figure for different sized logs. Measure the length and the diameter at the small end to get the board feet.

With a bandsaw mill, which has a very small kerf, small-scale sawyers like you or a neighbor can often overrun the scale. This means you have cut more lumber out of the log, more usable boards, than the conventional scale would suggest. The kerf is the width of the cutting blade. A circular sawmill blade often takes as much as a quarter inch per cut. That means every four cuts chews up an inch of lumber. A bandsaw generally takes less than 1/10 of an inch.

All lumber is bought and sold by the board foot. When professional foresters cruise timber they randomly sample the stand to figure how many board feet can be harvested. Pulpwood for paper and processing into chipboard is generally sold by the ton. Firewood is sold by the cord, which is a tightly stacked pile measuring 4 feet by 4 feet by 8 feet.

When you cut logs for milling, you generally want to cut them in multiples of 2 feet. The shortest thing you can use is usually 6 feet, although commercial mills limit themselves to 8-foot logs. Then you would cut 10 ft., 12, ft., 14 ft. and 16 ft. By using a bandsaw, we believe anything you can make 10 inches square is worth saving for lumber. Crooks, even slight ones, can easily take away 30 percent or more of a log. Try to cut your logs on the crook. It's much better to have two straight eights than a crooked sixteen. Plenty of good silvicultural books are available to get you up to speed on forestry and lumber if that is something you'd like to pursue.

Harvesting lumber on a small scale is possible, even though loggers will tell you otherwise. You can put some hefty boards down from the edge of a hay wagon, and with a couple of guys using cant hooks, roll that log up onto a hay wagon or low trailer. Another way is to put a cable under the log and pull it from the opposite side of the wagon with a tractor, creating a huge moving pulley that just rolls the log up.

We use our forks on the tractor, and as long as you do not have too big a log or get on a slope, or try to lift too high, you can lift some hefty logs onto a wagon or into a pickup bed. For real big ones you may want to get somebody with a backhoe to come over and load them for you. Although this is poor-boy, it works for small volumes. Of course, with a portable bandsaw mill, you can back right up to the log and mill it there.

One of the most fascinating stories I ever read was about a fellow up in one of the small New England states who had an 18-acre woodlot. He made a full-time living off that little tract. He had a woodworking shop. He examined each piece of wood, even each tree, with a value-added use in mind. The largest logs he milled into boards and made into furniture. Little 6-inch diameter pieces he put on a wood lathe and made rolling pins or lampstands.

When he cut a tree, he would cut the stump about 18 inches tall and then cut it again right at the ground level, where it flares out. He would dry that stump and then flip it over. That sanded and varnished flared end made a beautiful stool top. It just shows what some talent, creativity and tenacity can do.

Even within cuts we can do things, from a micromanagement standpoint, to increase diversity. For example, I often leave den trees and snags. Den trees encourage squirrels and other mammals that need a tree home. Snags serve as roosts for buzzards, crows and wild turkeys. It does something to my spirit to come upon that one-acre cut spot and see ten majestic wild turkeys roosting in a tree I left for them.

Some big old trees too become almost a part of the landscape. It seems almost irreverent to cut them down, especially when we've used them as landmarks. We have a tree we call "The Squirrel Tree." It's a huge shagbark hickory, very much in decline but still alive. Thirty years ago when I was a kid that tree always had squirrels crawling around on it. I spent many an afternoon sitting at its base waiting for a deer to come by, but being entertained by the constant chatter of a dozen squirrels.

Someday that tree will come down in a windstorm or be struck by lightning and burn to the ground and I will regret that my grandchildren will only be able to see a mound on the ground and hear me say: “That is where the squirrel tree used to be.” The landscape diversity becomes our own diversity, our own distinctive character. How dreadful to see every tree as just so many board feet of lumber, or to see every field as so many corn plants.

In short, diversifying the landscape is the foundation for diversifying everything else and to being able to have a balanced, stable farm. To keep predators well fed so they do not eat your chickens takes healthy forestal and riparian areas growing small mammals and birds. It all works together in a great harmonious chorus. Any farmer will be ahead by creating diversity everywhere possible.

Photo 23-4. *Riparian areas offer wonderful biodversity opportunities, but should be protected in order to be fully functional.*

Water

You can never have too much water, except in a flood. Building in redundancy for on-farm water systems is a good objective. Obviously this subject is too broad to be comprehensively covered in one chapter, but here are some basic guidelines. Three primary water sources exist: wells, springs and rainwater runoff.

Every area has unique properties. I would always have a well site dowsed before digging. Every community has a dowser; ask around until you find one with a good track record. Preferably locate the well on a high spot to take advantage of gravity for whatever you do with the pumped water. The higher you start, the farther you can push.

In remote areas, solar pumping systems are now cost effective and very dependable, although they usually are competitive only in low volumes. That is why most solar pumping systems use a large storage tank.

One of the primary principles about transporting water is that a dependable high volume, high-pressure system requires no storage while a low volume, low-pressure system requires high storage. You can either pay for storage or pressure. You need not pay for both.

For example, you can water 100 cows out of a 3-gallon bucket if the water flows in faster than a cow can drink. But if you have a slow flow you need a big tank to water all those cows. If the water comes in slower than they can drink, they will tear up the tank and each other trying to satisfy their thirst.

Your resources will dictate whether you want to put money in pressure and volume, or storage. One of the most common mistakes, in my opinion, is putting too much money in pressure and volume. For example, using our illustration of 100 cows, we would need a system capable of delivering at least 15 gallons per minute if the drinking trough is less than 50 gallons. But if we went to a 350-gallon drinking trough, we would do fine at a delivery rate of only 1 gallon per minute.

Those 100 cows will drink 1,000-1,500 gallons per day. At 1 gpm we're getting 60 gallons per hour and 1,440 gallons per day. A 350 gallon trough costs only $150. We can use small pipe or perhaps gravity to deliver that 1 gpm. But a system delivering 15 gpm from an equivalent distance will require a sophisticated pressure tank and pump, large and expensive pipe and a large volume, dependable source at the pump. Granted, the large tank is not as easy to move from paddock to paddock as the small tank, but for the additional capital and maintenance cost, you can scoot a large tank around for awhile.

Let gravity work with you instead of against you. It is always easier to drain water than push water. If your well cannot be located at the top of a hill, then put in a reservoir at the top of a hill. Run a pipe up to it and then run the water line network downhill, like a spider web, from the reservoir. That way you can slowly fill the reservoir 24 hours a day but if you need a large volume of water, as in a mid-afternoon watering for 100 cows, the reservoir's storage can deliver the volume. Gravity gives you pressure without sophisticated pumping systems. This means you can use much smaller and cheaper pipe to deliver the same volume of water.

A reservoir can be something as simple as a used underground fuel storage tank from a filling station. Many of those are 5,000 gallons or more. Just set it on some rot-resistant timbers to keep it

Photo 24-1. *A simple float valve on a small trough gives water for chickens, cattle or hog watering nipples. Clean, fresh, pressurized water on the whole farm creates all sorts of management and production opportunities.*

off the ground, plant some ivy around it that will grow up over it and keep it cool, plumb in a float valve and you're in business. It can be a hole in the ground like a swimming pool. I heard of a guy once who built one out of woven wire fence. He made a ring of fence and then covered the ground with old hay or straw. Then he used a big piece of seamless plastic to form the tank. As the water slowly filled the plastic, he kept adding straw between the plastic and the fence in order to give a soft wall and keep from poking a hole in the plastic. When it was full he had a 30-ft. diameter, 4-ft. high cistern of water.

Of course, if you're on flat land, you may need to put a few poles in the ground and lift the tank up on a pedestal in order to get enough pressure. If the land is that flat, however, you may just as well push the water directly from the well.

The second primary source of water is a spring. Whole books about spring development have been written and they can be checked out of a library. The main thing about using a spring is to not impede

Photo 24-2. *Nothing works like gravity. From nearly a mile away, without any pumps, the pressure is enormous.*

the natural flow of water — have your outlet pipe at the same elevation that the water is coming out of the rock. The two types of springs are those that bubble up from the ground and those that come out of a rock ledge.

In either case, you want to eliminate surface water contamination and blockade the free egress of the spring water by putting it through a pipe. Once the water is in a pipe, you can send it anywhere. If the spring is strong enough, you can pump directly out of a little box. If the flow is low (anything less than 5 gallons per minute) you would want to run it into a reservoir. Multiplying the value of your reservoir by building a heavily insulated structure around it can be quite beneficial. Depending on the flow and how cool you can keep it inside, it can be a quasi-root cellar.

Generally you want to excavate carefully around the spring to expose the water seams, then pour a concrete wall with a pipe in it. Placing the pipe too high can create abnormal water pressure behind the wall, encouraging the water to find an easier access around the wall. Make sure the pipe is at the water's entry level. Place a screen over the pipe to keep critters out. Remember, though, that you will never be able to completely seal off the system from little critters. They live in every spring.

Then you can build a little roof over the spring to keep out surface water. A little diversion ditch above the spring can help route surface runoff around it.

The third water source is rainwater. Although we do not have a cistern on our farm, I have a great appreciation for them. We are blessed with a very strong well and have built lot of ponds, so building a cistern has not been a high priority.

The biggest mistake when installing a cistern is not making it big enough. Since there are 231 cubic inches in a gallon, a one-inch rain on the roof of a 1,000 square foot building is more than 600 gallons of water. That is not a large building, either. When you figure a typical shed or barn, or even a normal house, the volume is staggering.

The reason cisterns often have a bad reputation, though, is because they go dry because they are too small. It doesn't cost much more to add another few thousand gallons during construction, and the increased storage allows you to capture all the runoff. The goal is to never have the cistern full and running over.

This brings us to ponds, which capture surface water runoff and store it for later use. I trust the wisdom of local excavators when it comes to pond building. I select the site and let a good track loader operator decide how best to situate the dam. Operators who have built dozens have a great feel for what is do-able. They know what kind of material will hold best and how to slope it for maximum efficiency.

If you trust the government, you can have a technician from

Photo 24-3. *Ponds are one of a farm's most important assets in order to capture and store water. This pond is crystal clear—you may be able to see the leaves on the bottom.*

Photo 24-4. *A 500,000 gallon pond at the base of a high ravine offers year-round water to our farm.*

the Natural Resources Conservation Service come out and design the pond. In my experience, though, these ponds are overbuilt. If you find a technician who advocates sealing it with pig wallows, let me know.

A pond can be something as simple as digging a hole in a wet area with a backhoe to let the water seep in to something as extravagant as a gated spillway. Some areas are more conducive to ponds than others because of geology, and local wisdom can help guide you in such circumstances.

Do not allow cattle access to the pond. They push the sides in, devegetate the sides and defecate in it, which destroys the water quality both for fish and for drinking. If a pond leaks, put hogs in it and let them wallow in that mud. They will pack that bottom and seal it up. The water level will be where the leak is, so no need to drain it.

We put in a pond recently in the mountain where shale seams often cause problems. Sure enough, the pond only held about one

Photo 24-5. *A floatee keeps the water coming into the system about 18 inches below the pond surface. This is the best quality water.*

foot of water; it was supposed to hold about six. We put pigs in there for about two weeks and they made quite a wallow. We fed them corn right down at the water line to encourage them to tromp it all up. The next time it rained the pond filled right up as pretty as could be and has been holding water fine ever since. Old-timers have a lot of wisdom if we will just sit still long enough to listen.

Locate a pond as high up on a hill as possible, rather than clear down in the bottom. Although you lose runoff area locating it high up on a hill, you gain gravity flow. The best water in a pond is about 12-18 inches below the surface. For watering livestock, pull the water from this stratum.

We have a pond at the foot of a long valley that feeds our

Photo 24-6. *A pipe siphons the water through two float valves into a 600-gallon tank. An exit at the bottom supplies pressurized water to nearly three miles of 3/4-inch pipe for the farm. The tank acts as a ballast; it also means the system runs on pushed water rather than pulled water.*

livestock system. We can also use it for garden irrigation. A siphon pokes through a piece of Styrofoam that floats on the surface. It can rise and fall with the water level. The siphon goes down to a 600-gallon tank that acts as a reservoir for the system. This storage tank acts not only as a ballast to cushion the system from high usage periods, but also gives several feet of head pressure at the start of the system, to push the water into the system. I think it is unwise to run a multi-mile system like ours on a siphon; better to have it push than suck. The storage tank allows us some ballast in case we pull water faster than it can suck in the siphon. This system gravity feeds the entire farm. The 80-psi ¾-inch pipe runs on top of the ground and every 200 feet is a 'T' with a little 99-cent hardware store plastic ball valve. We run this water line up some high hills, but as long as we do not go higher than the reservoir tank, we have water.

Photo 24-7. *Every 200 feet a 'T' and simple hardware store plastic ball valve offer pressurized water. In-line shut-offs facilitate trouble-shooting. The hose lies on top of the ground underneath permanent internal electric fencing. For winter, we open the valves and let the system drain.*

Streams and rivers require similar care as ponds inasmuch as they should be protected from stock and the water needs to be pulled out of them somehow. A water ram pump can be used in flowing water situations where plenty can be wasted. These pumps use the excess water to pump smaller amounts of water at high pressure.

A tool we've used quite effectively is a high-volume, low-pressure 12-volt marine bilge pump. We have a couple of 3,500 gallon per hour submersibles and they are great for moving high volumes of water short distances. They will not pump water far or high because they are designed to pump water out of the bilge and over the side of a boat. We just attach a 50-foot piece of 1¼-inch plastic pipe to the outlet and run it over the side of the pond or creek to a water trough. It can fill a 500-gallon trough in just a few minutes while we check the livestock or put up a cross fence.

A pair of alligator clamps to connect it to a deep cycle battery works fine for powering the pump. About the size of a softball, the pump is completely submersible. We just drop it in the pond and watch the water flow. This pump works great for any application where you are trying to move a lot of water fast under low pressure. We've used it to pump out barrels of water, start siphons, and pump water off the swimming pool cover in the spring.

With the battery as a power source, it is as good as a well for pumping water where you have no source of power. A friend turned me onto this pump when he was trying to mix concrete in a remote area for a cottage. He needed to get water up to the site and this worked like a charm.

Landscaping for better water utilization is a good technique, and again one that the permaculture folks do well. Digging swales just off contour to duct surface water either into a pond or just keep it from running down the side of the hill can be used to great benefit, especially in arid areas. This technique is used as well in books on spring development, to increase the catchment of the area. A small pond the size of a swimming pool, for example, can receive double the water if you mound up some wings that flare out kind of like the sides of a funnel, to catch more surface runoff and duct it into the basin. Israelis now routinely hill up a little dam on the downhill side of tree seedlings and this increases water retention in that spot by 300 percent. In a low rainfall area, this is the difference between death and abundance.

One thing I've done to help decide on landscaping techniques to get better control of water is to don the raingear during a downpour and go for a walk. This is especially effective on a road. Sometimes until you actually see the water running you can't be quite sure about how to control it. In the spring right after a snow melt is also a good time.

We've built water bars across ditches, or even a little diversion on a cow path, just to get water slowed down and the volume reduced. Water destroys when volume and velocity build up. Essentially landscaping should reduce the amount of water that flows in any one place at a time, and slow down what water does run.

This is especially important on roads. Keep them water barred (diversion mounds running perpendicular to the road that channel water off) to knock off the water frequently. Gravel as much as possible to keep the raindrops from hitting raw dirt. Have the gravel truck put a 30-inch block in the back of the truck to spread the gravel just in the tracks. You can get at least a third more footage covered that way. Then the middle can grow up in grass.

Overall, we need to look at all the places where water can be a liability and turn those into assets. Whatever you can do to gain more control over the water that hits your farm, the better. Too often water is the limiting factor for production and profit. Capture all the water and put it to use.

Letting Animals Do the Work

"How do you get it all done?" people ask. My response: "Oh, I don't do anything. I'm just an orchestra conductor making sure all the animals are where they are supposed to be at the right time. They do all the work."

While that may be a slight stretch, it certainly is more true than false. Every animal has abilities that can accomplish tasks on the farm to reduce fuel, labor and machinery expenses. The beauty is that animals don't sue when disgruntled, don't require minimum wage, don't need replacement parts, don't need their oil changed and don't even require pension plans. And they're just plain fun. Utilizing animals as our labor force completely changes the farm's economics.

For example, large scale composting is a labor-intensive and machinery-dependent chore. We generally make more than 200 cubic yards of compost per year, most of it generated as a winter bedding pack in the cattle hay feeding shed. Years ago, we used to clean this packed manure-chip-sawdust-straw material out, load it on a PTO-powered manure spreader and build long windrow-type compost piles.

Barn bedding should always be converted from anaerobic to aerobic before field application in order to fully capture nutrients. The composting process pre-digests the material, virtually eliminates odors (nutrient vaporization), sanitizes the material through the heat-

Photo 25-1. *"Pigaerators" inject oxygen into cattle bedding, creating aerobic compost through the mixing action. This equipment appreciates in value, doesn't need spare parts or oil changes, and comes with a free operator.*

ing process and stabilizes the nutrients.

This bedding clean-out and compost pile building is no easy task, especially after animals have packed it. For that matter, even backyard compost pile turning is no picnic.

What we do is add grain to the bedding during the winter and when the cows come out in the spring, we turn in some pigs. In 8 weeks two 200-lb. pigs will turn 75 cubic yards of material on half a ton of grain. They are pure joy to watch. They seek out the fermented grain kernels, digging and mixing the bedding, tearing it all apart and creating the finest compost imaginable — far better than that built by machine. They turn it multiple times and will dig down to 3 feet. And when they are done, we harvest them for pork. Pretty simple retirement plan.

We have wonderful compost without ever turning on a tractor or spending any time. The pigs produce fabulous meat, are fun to watch, and do not smell. Most importantly, we haven't broken a sweat. And when we do clean it out with the front end loader, the compost is all torn and decomposed into nice loamy material. Digging it out is easy on the machinery, as is the spreading.

The same procedure can be utilized in a horse stall, underneath rabbit cages, or on a large garden compost pile. Just inject some grain, put some gates around the pile, and watch the animals enjoy doing your work. It'll be more delightful than Mark Twain's story of Tom Sawyer getting appreciative tokens from the neighborhood boys for the privilege of whitewashing the fence.

We even use pigs to do heavy tillage. Electric fence works great with hogs, and they can be rotated from area to area. We use two strands of wire at about 4 inches and 12 inches. A 200-lb. hog will till 100 square feet of ground each day on 5 pounds of grain. Corn and small grain both grow fine behind hogs' tillage. It is so simple and enjoyable, the lucrative monetary part of it seems incidental.

We have a 12 foot x 20 foot "Tenderloin Taxi" that we pull around areas we want tilled, and four pigs in the taxi do a great job. We used it last winter to clean up after Daniel's mangel harvest. He grew a 3,600 square foot patch of mangels for winter supplemental feed for his rabbits. A close relative of the sugar beet, these tubers are impossible to harvest completely. You always miss a few. By pulling the tenderloin taxi across the top after harvesting, the pigs tilled up the area and harvested the rest of the crop that escaped us.

In a related use, what began as another experiment to convert forest to pasture is turning into additional research and development of sustainable self-harvesting hog production. In this scheme, we identified a gentle slope in the forest that we would like to convert to open land — initially, for pasture.

We harvested all the trees as firewood and sawlogs. Then we partitioned these two acres off into about 8 paddocks with electric fence and put hogs in — real honest-to-goodness "bush hogs." The pigs tilled up the area. We try to shift paddocks at least every two weeks. A couple of days before we move the hogs, we go in with a little handheld cyclone seeder and spin on small grain. The pigs eat some and tread most of it in.

We can let that paddock rest for 14 weeks before coming back to it. By that time the grain matures and we turn the hogs back in. They hog down the crop, self-feeding on the grain. In addition, they till up the area again and we can plant again to keep the cycle going. This land is extremely rough, scattered with branches, stumps and rocks.

The second year, the land begins to open up a bit. The pigs have killed most of the mountain laurel and little saplings. It is now ready to grow corn. After the corn gets to the milk stage, we turn the pigs in and let them hog that down. We have experimented with corn by planting it with a sharp stick on wide spacings, as well as densely broadcast. The broadcast does not yield the ears, but the biomass is incredible. The hogs eat ears, cobs, leaves, stalks — everything. Each seems to work fine.

Our biggest problem has been germination during dry times. Since we are planting things all season long, we hit some plantings perfectly ahead of a rain and other times we plant and get no germination because of drought. This year we built a pond to gravity-feed water into the entire two acres and are looking forward to being able to add water when necessary to ensure germination.

We keep a 12-hole self-feeder with the hogs at all times containing corn, roasted soybeans, plus small grain. When the hogs have all the corn plants they want, they go weeks without touching anything in the self-feeder. That means they are fully satisfied with the standing crop. We tried unsuccessfully last year (because of the drought) to refine the feeding program through diversifying the plants offered to the pigs. Adding pole beans to the corn hole gives a high protein with the starch. Since this is rough ground, some areas do not germinate. Those can be filled in with zucchini, which pigs dearly love.

This is an ongoing project, but it has been truly exciting thus far. The paddocks are gateless. The pigs move from paddock to paddock easily if we just take a couple of rocks and put them on the wire about 8 feet apart. We make sure the pigs get hungry by running out of feed for 24 hours. Then we just throw a bucket of corn into the next paddock and they come running across the wires. If the weather is real hot, we throw a couple of buckets of water into the next paddock and that entices them just as well. After the pigs move, we pick up the two rocks, the wire flips back to its original position, and we're all set.

Each paddock has a T-post in the corner with a hog nipple attached to a pipe sleeve with a large set screw on it to hold it from slipping down the T-post once we have it at the desired spot. Gravity-fed water down a pipe from a pond supplies pressurized water to the nipple. Enough saplings continue to grow to provide enough shade for the pigs.

This is an ongoing project and we don't claim to be experts on it yet. But for sure we have not begun to explore all the possibilities. When I was a kid, we grew a patch of Jerusalem artichokes for a pair of hogs. Any tuber is perfect for a hog. Our next objective is to actually get our production of feed to the point that we can quit buying in grains from outside and let the hogs self feed on fresh-grown material. Using them as the fertilizer and tiller certainly makes this an economical goal.

In addition, we use a portable pen with little pigs in the winter to keep the hoophouse bedding from capping. The laying chickens, when they have fewer than five square feet per bird, often apply more manure than they can adequately incorporate into the bedding with their scratching. Sometimes, for one reason or another, the bedding gets damp on top and that adds to the capping problem. Large poultry factories use rototillers and special equipment to deal with this. But with a little pig pen that we walk down through the houses, the pigs root deeper than the chickens and break up the cap. Then the chickens love to come behind the hogs and find all the newly exposed treats.

Pigs can also be used in place of a rototiller in the garden, but this must be done judiciously. We built a 6-foot x 10-foot "Porkovator" for two hogs that we could walk down through the garden. When all the soil conditions are perfect, it works wonderfully. But if it's too dry or too wet, they can do more damage than good. I don't believe they will substitute for a tiller or spader in large scale gardening simply because when it is time to put in the vegetables you can't get the pigs over enough ground fast enough to make it logistically feasible. But certainly for an initial sod tilling, hogs are superb. As long as you move them about frequently, they will never smell. They are the most delightful creatures you can imagine.

Photo 25-2. *Betsy and Amanda Womack pet the pigs in their "Porkovator," a portable pig pen moved routinely to till up garden spots. The hogs till, fertilize, and provide enjoyment.*

Recently we've followed the lead of a friend near Bristol, VA who goes out after Halloween and salvages all the leftover pumpkins. Stores are happy to give them away and they almost provide a full ration for the hogs. This gives free feed for a month.

We now take the spoiled food from the area food bank that would otherwise end up in the landfill. The volume fluctuates dramatically throughout the year, but generally we get a huge pickup load every week. It will be as diverse as cracked eggs to donuts to sweet potatoes. We save hundreds of dollars in feed and the pigs seem to relish the different foodstuffs.

The main reason farmers have had to get bigger to stay in business is because most of them are essentially in the materials handling business. Figure out how much soil you have to move to till an acre, and you will see that we are indeed in the materials handling business.

Materials handling is cheaper by the million, cheaper by the ton. Operating a 3-yard front-end loader is far more efficient than operating a 1-yard bucket. Moving grain with a 6-inch auger is far more efficient than running a 3-inch. Hauling 20-ton lots is far more efficient than hauling pickup loads.

Obviously, if you're in the materials handling business, getting bigger is essential to remain competitive. But if we let animals do the work, this principle does not hold true. Harnessing animals enables profitability to be size neutral. Instead of using depreciating tractors, we're using appreciating equipment. That is the way to profitability. And with hog prices low, this is real cheap pig iron. Buying a tractor for 30 cents a pound is not a bad deal.

We do not attempt to farrow hogs, because then we'd be in the hog business. We do not want to really be in the hog business. Sows and boars depreciate — let someone else own them. We just buy weaned hogs, so all our money is in appreciating stock. We are in the profit business, not the hog business.

If we don't have to pay for the machinery, we can make a profit on the first pound of material produced. Ireland agriculturists have now concluded that unless you put more than 1,000 hours a year on a tractor, it is more economical to contract out tractor work than it is to own one. This illustrates the tremendous overhead costs of the capital expenses required to be in the materials handling business. If your profit margins are high enough through value adding and retail marketing, you can afford to own a tractor for fewer hours.

Now we'll turn our attention to chickens. In the chapter on

Photo 25-3. *The "Raken House" utilizes aggressive laying hens underneath rabbit cages to incorporate carbonaceous bedding material and the rabbit urine, producing a magnificent compost. This eliminates odor, which keeps the rabbits above healthy, and builds tons of compost. Bedding only needs to be cleaned out once a year.*

"stacking" we deal with these models as well, but here we're putting a different angle on them. The chickens under the rabbits in the "Raken" house provide the benefit to the rabbits of aerating the bedding to eliminate odors. Any commercial rabbitry smells terrible. The rabbits live with their noses stuck into the equivalent of a bleach bottle all their lives, from the noxious ammonia vapors coming up from their urine. But the chickens incorporating this into carbonaceous bedding eliminates all that, which keeps the rabbits above healthy. With their scratching, the birds not only build compost out of the rabbit droppings, but also keep the rabbits healthy, which eliminates the need to add medication to the water. Virtually all commercial rabbitries medicate the water routinely to deal with disease problems.

The "Eggmobile" that follows our cattle is another example

of the animals doing the work. While other cattle producers are out getting kicked in the shins trying to put cattle through a headgate so they can put Ivomectrin down their backs (which makes the meat so bad it kills all the bugs), we just move the chickens around behind the cattle and collect $4,000 worth of eggs per year as a byproduct of the pasture sanitation program.

The chickens eat the fly larvae, turning a liability into an asset, and incorporate the cow paddies into the soil to stimulate nutrient cycling. This is often done by high-class cattle operations with a flex-tine pasture harrow. But the chickens do this process just fine and lay eggs to boot. Instead of being a John Deere jockey dragging cow pies to the sounds and smell of diesel engines, we can listen to the happy cackling of laying hens.

This cow paddy breakup also reduces the repugnancy zones

Photo 25-4. *The Polyface "Eggmobile" with its 410-hen complement of workers. The hens fellow the cattle in the grazing cycle, scratching through cow paddies and eating fly larvae, incorporating the manure for better nutrient cycling, and eating hundreds of pounds of insects. Moved daily to a new area, the birds sanitize the paddocks for livestock, debug the field, and produce $40 worth of eggs PER DAY.*

(the ungrazed area around manure patties) so that cattle will graze more evenly the next time they come back to the field. Using chickens around any kind of cattle is a positive mix. Feedlots and dairies can benefit immeasurably by having chickens coming along to eat fly larvae, harvest undigested feedstuffs, and scratch up cow pies.

Andy Lee, in his book *Chicken Tractor,* explains how he uses chickens as tillers in the garden. Using a smaller pen than our field pens, he moves the birds around the garden and they till it, eat weeds, eat bug larvae and add fertilizer.

I do not believe we've scratched the surface — pun intended — with what chickens or other poultry can do in horticultural and vegetable production operations. Can you imagine turning loose 100 turkeys on a grasshopper-infested crop field? A great idea is turning turkeys into a cornfield once the corn is about 2 feet tall. The turkeys keep it debugged, side-dressed with fertilizer, and use the corn plants for shade. The corn is too big for the turkeys to destroy, but the birds readily eat all grass and weeds. It's a wonderful combination. Although we have not personally tried it, the literature is full of good stories about weeder geese in applications from vineyards to strawberries.

In arid irrigated areas where bugs are a problem on hot peppers and other specialty crops, chickens afford a wonderful pest control. Nothing works better than chickens to eat dropped fruit, eliminating the over-wintering bugs at the same time. Whenever chickens get access to trees, they constantly scratch around the base looking for bugs. Orchardists routinely fight worms, fungus and insects around the base of trees. Chickens disturb this area looking for these goodies and thereby break the pests' cycles.

What about predators? The first defense is to keep the chickens moving. That keeps the predators off guard. Second, provide predators plenty of food by having wild areas and healthy populations of small mammals in thick grass or leaf duff. Third, you may need to use guard dogs or guard donkeys.

Now we'll move to cattle. By far the most efficient and cost-effective way to harvest grass is with the four-legged mowing machine. But a mower, to be useful, needs two things:

1. An on/off switch so it can be started and stopped.
2. Maneuverability (a steering wheel on a riding model, or a handle on a push type).

Portable electric fencing is the switch and steering wheel for the four-legged mowing machine. We can put the animals where we want them when we want them to be there, in whatever size or shaped area we want.

For example, in the spring before we turn the bull in to begin breeding cows, we use him as a lawn mower to tidy up around the

Photo 25-5. *"Alvaredo" the Brahman bull mows around the house and outbuildings during early spring before breeding. He does a better job than any weed whacker. Tiny electric fence paddocks contain this big daddy.*

shed, the other outbuildings, and the yard. We'll just give him a little place, maybe 100-200 square yards, and leave him for a couple of days. He loves the fresh grass, we don't need to feed him mechanically harvested hay, and it keeps the lawnmower and weed whacker silent. As a side benefit, he fertilizes. Actually, up around the house the fertilizer can make life interesting, but the benefits outweigh a few black pumpkin pies lying around. Putting up the electric fence takes just a few minutes and he's as controlled as he would be if we had a concrete barrier around him. Visitors love to go up close to him — outside the fence — and touch his nose. It's great entertainment. When the bull is with the cows, we use the milk cow in the same way.

On a grander scale, we use the cow herd to mow along the roadside ditches. When we are in a paddock adjacent to the road, we can open the gate and take a magazine along to read while we let the cows graze the roadside. A person on either end to hold them in case a vehicle comes works real well. Obviously you wouldn't want to do this on a heavily traveled road, but there are thousands of miles of dirt roads in this country that would provide many grazing days for cattle — and a bunch of highway department mowing machines could be parked as a result.

I spoke at a grass and range conference in Mexico a couple of years ago and on the drive from Monterey to Victoria was impressed about several things. All along the main highway milk cows were staked out eating the grass. In the evening, the farmer would go out and milk the cow and move her stake. It was a wonderful way to keep the road edges mowed and provide food.

Secondly, while I was in Victoria, I saw the expansive park in which the folks there take great pride, but staked in the park were some milk cows. My host said that six folks in the city had milk cows in the park and were responsible for keeping it mowed. They lived right in the city and yet had their own milk.

To me, that makes our batwing mowing machines and the countless taxes collected to run them seem like Neanderthal policy. How elegantly simple to use animals in this way. All it takes is some control, whether by electric fence or a stake, and we have a mowing machine that produces food instead of consuming parts and petroleum, not to mention the difference in the exhaust.

We use cattle to harvest as much of their own feed as possible through short duration, high density grazing. This past winter we fed hay for only 68 days; most farmers in our area fed hay for 140 days. It's typical for farmers to brag about how much hay they make; I like to brag about how much hay we did *not* make.

In the spring, we let the cattle graze the entire farm — moving them fast from paddock to paddock. This opens the whole farm as a grazing pool while the grass is growing slowly. We graze hayfields and everything. An unbelievable number of farmers do not allow the cattle into their hayfields or alfalfa or small grain, but continue feeding hay for an extra 4-6 weeks. Right across the fence is a beautiful stand of forage a foot high that could be self-harvested by the cows.

Research shows that harvested forage costs at least three times more per ton of Total Digestible Nutrients (TDN) than does the same material harvested by the grazing animal. It is beyond me why so many farmers insist on mechanically harvesting all that material just so they can haul it to the cows. The cows would gladly harvest that forage.

When we can reduce stored feed requirements to just a couple of months, it completely changes the labor requirement during the spring and summer because we are not having to mechanically harvest forage. But when cattle are free to graze areas continuously so that pasture productivity is 25 percent of what it should be, the shortfall needs to be made up with mechanically harvested and petroleum fertilized feed.

By controlling the animals to capture natural grass growth patterns, the profitability of the farm completely changes. The big expense items like diesel fuel, machinery, parts and labor become practically nonexistent. The aeration accomplished by high density grazing keeps hayfields thick instead of clumpy, eliminating the need to overseed or refurbish old fields.

The cattle hooves are tillage tools that can be used detrimentally or beneficially, depending on placing, timing and density. Overseeding clover is necessary only when continuous selective grazing weakens legumes and favors unpalatable forage species. Natural succession toward legumes and palatable species will kick in under proper grazing management. *Salad Bar Beef* details grazing management as a tool.

Grazing built the fertile prairies, and grazing can rebuild the soil. No plows, no combines, no monocultures. Herbivores and perennials are a winning combination if managed properly.

Perhaps someone is reading this chapter wondering why the most common working animals have been ignored: horses and oxen. A couple of reasons come to mind.

(1) That's too normal and I'm an abnormal kind of guy.

(2) I do not know much about horses and oxen.

My apologies, all you horse lovers, but I admit it: I'm scared of horses. We had some ponies when we were kids, and they were the orneriest, boniest-backed critters I've ever encountered. They never went where we wanted them to and when they finally got tired of our persistence they just remedied that problem by bucking until we hit the dirt.

I've been told that ponies are much worse than horses. That may be true, but in my memory, they sure look a lot alike. When I turn the key off on that tractor, it just sits. But there's something unnerving to me about routinely being inches away from a one-ton power source that can think. That scares me. I know, people used horses long before tractors came along. But I didn't grow up then. I grew up now.

Call it foolish, call it ignorant, call it unenvironmental. I just do not have any hankering at all to place myself around and under and beside a thing that can flick me against a wall like a fly. I love to watch horses race; I think they are beautiful creatures. I pet them when I get a chance, and get teary-eyed when I see a perfectly matched six-horse hitch in parades. But there's a big difference between enjoying their power and beauty from a set of bleachers, and sidling up under one to pick up his foot and take a look. No, my friend, not this boy.

My hat is off to anyone who works horses or oxen. I think they have a wonderful place on small farms. For cultivation, they are wonderful. I'll never forget one morning going to see a man who was picking field corn. The morning sun was just coming up to melt the frost, and here was a man and his team of horses out in a cornfield. Solitude. All you could hear was the horses breathing, and about once every two or three minutes the farmer would mutter: "Get up . . . Whoa." The horses would move up a few feet and then stop. It was poetry. I marveled at the control, the elegance of the work being done and the obedience of these big animals.

Perhaps my children will bring these gentle giants into my life. Perhaps not. But I am ignorant and scared and it won't do any of us any good for me to put on an act. If you have a desire to work with horses, by all means pursue it. Plenty of learning opportunities exist because animal power is definitely making a comeback.

(3) My people resources know how to fix machines, not horses. That is the reality of my neighborhood and family. That affects my decisions, and right now it means that fixing the tractor is relatively simple. Besides, tractors come with warranties.

(4) I would need a tractor anyway. I've often said that when they figure out how to put a PTO shaft and a front-end loader on a horse, I'll be ready to buy one. We move a lot of material and I can't see owning horses when I need a tractor anyway. Horses are not cheap to own, despite what some fans say. From a purely economic standpoint, it is not a black and white issue. Each has positives and negatives. Do not be overly romanced by some horse or ox enthusiast and lose your head over what is reasonable.

A forecart with ground-driven PTO, which is not nearly as good as the one on a tractor, will run $5,000-$8,000. In addition, you have the horses, and the horses need to be worked and fed, even when you don't need work done. At this time in history, and for my experience and temperament, I just am not into horse power. But for those of you who are, please pursue your dreams. Perhaps even I will see the light someday.

I'm sure many permutations on this whole animal concept exist, but I hope this discussion stimulates your creative juices. Sheep in orchards, goats in weeds, llamas in sheep, dogs herding sheep: the list is endless. Don't forget about sewage-feeding fish and manure-eating, humus-building earthworms. If we harness the innate characteristics of animals, we can reduce our workload and greatly increase our bottom line.

Livestock Sanitation

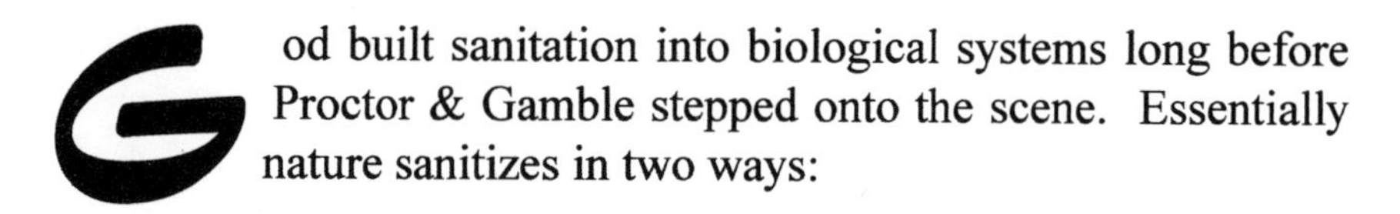

God built sanitation into biological systems long before Proctor & Gamble stepped onto the scene. Essentially nature sanitizes in two ways:

1. Rest and sunshine
2. Decomposition.

Pathogenic organisms thrive in conditions that encourage:

1. Exclusion of direct sunlight
2. Dampness
3. Constant host access
4. Single-host concentration
5. Improper decomposition

Multitudes of good organisms exist to control bad ones, and these need to be encouraged in our livestock and plant production models. If you will put attention on these five problem areas when designing any livestock facility, you will virtually eliminate disease and odors — the things that make animal buildings unpleasant.

Let's take a look at these five elements one at a time. The greatest sanitizer in nature is sunlight — especially direct sunlight.

This means that facilities that house livestock should incorporate plenty of sunshine into design. Skylights and open eastern sheds encourage sunlight penetration to the floor and onto the animals. Translucent fiberglass panels for skylighting should be clear rather than green-tinted. In addition, skylights should run all the way from the peak to the eaves rather than inlaid between metal roofing. When rainwater comes off fiberglass or plastic type material, it is caustic to metal and will rust roofing quickly. To prevent that, we run skylights clear to the eaves on any building in which we install them.

The second item, dampness, is encouraged two ways. First, if air circulation, or ventilation, is lacking, that dank, musty odor will hang in the air. Most animals can take cold, but they cannot take drafts. A shed closed to the windward side but open to the front may seem cold, but the animals can snuggle into the wall and stay quite comfortable. Any crude pen with roof and wall on the windward side will work in most situations. The combination of drafts and dampness is a sure animal killer.

Since animal excretions are extremely damp, it is necessary to absorb them with dry bedding. Bedding should be stored under roof to keep the absorptive capacity from being filled with water. Straw used for bedding should be rained on a couple of times before being baled in order to break that shiny outer water-repellant coating. Rain makes it soft and absorptive.

One of the single biggest pathogen stimulants is constant access to a host. This is especially true of parasites, and is probably the single biggest problem to homesteaders with limited acreage for stock. Most parasite populations only stay virulent enough to pose a danger for about two weeks without the host animal. If the stock can be rotated from spot to spot pathogens stand a harder chance of getting a virulent foothold.

Whether it's a baby calf in the backyard or a flock of six laying chickens, this denial of host access is extremely important. Throughout history, this was a function of the shepherd — leading the flock to fresh, rested, hygienic pastures. For extremely small numbers of animals, portable pens serve well. For large animals like cows and horses, portable electric fence is cheap, easy and highly functional.

Lest you miss the point of this discussion, let me make it very plain: you probably do not need a barn. And if you do, it probably should not look like anything your local barn builder has in mind. And if you have the terrible misfortune of acquiring a piece of land with a bank barn on it, you probably should tear it down as soon as possible to build something more animal-friendly. Bank barns are the darkest, dampest, coldest, most inefficient buildings you can imagine. While some may argue that they were extremely functional for what they were built for, I would even take issue with that. You can't get into them with machinery to apply carbon or clean out. The low ceilings make deep bedding difficult to achieve. That lumber can much better be used in a simple pole construction or some other more functional building. Bank barns are nostalgic and handsome to look at, but they are perhaps the most dysfunctional building ever contrived for agriculture.

Especially for small groups of animals, portable facilities and simple hoophouses for winter housing work wonderfully well.

The fourth item has to do with single hosts. The more diversity we can plug into the livestock system, the more pathogens are held in check. Most pathogens are species-specific. Gardeners have long utilized multi-cropping systems to thwart insects. If every other plant is unpalatable, the pest expends energy traveling and being disappointed — imagine every other spoonful of food being inedible.

A corollary principle is multi-species symbiosis. In nature, birds follow herbivores, cleaning up after them. The egret perched on the rhino's nose is no accident. Multi-species husbandry is a good way to combat pathogens.

Finally, improper decomposition is perhaps the single biggest factor regarding pathogen populations. Agronomists call this the "decay cycle." The more active the decay cycle, the more balanced and healthy the soil and plants.

Organic matter feeds the decay cycle. A clay chicken yard, for example, destroys the decay cycle, as does a mud calf yard or non-grassed horse stable. Unvegetated areas (excepting areas where a green manure has just been tilled in) do not have active decay cycles. An active cycle requires proper proportions of:

1. Air — maintained by spaces between soil particles through earthworm activity and dead root tunnels. Soil "fluff" or tilth is superior to hard, compacted soil.

2. Water — penetration and retention are stimulated by a thick sod cover and high organic matter sponginess. Moisture provides a movement medium for microbes that travel in a microscopic swamp, sloshing around from spot to spot. In a housing situation, this is usually amply provided by urine.

3. Carbon — cellulose-rich material like leaves, straw, sawdust, wood chips, etc. Mature vegetable matter as opposed to green and tender vegetable matter like lawn clippings or kitchen wastes.

4. Nitrogen — supplied by immature vegetable matter like lawn clippings and kitchen wastes, but more normally provided by animal manures.

5. Microbes — usually living in the soil, but can be encouraged with innoculants.

Heavy animal compaction hardens the soil, pressing out the air and shutting down decomposition. If the manure load is too heavy, nitrogen overloads and the area becomes toxic. Pulsing of the vegetation during rest periods (when we keep animals off an area) adds carbon and allows earthworms to aerate the soil and bring up necessary minerals. Earthworm tunnels encourage water penetration and provide channels for good capillary action during dry periods.

Sanitation does not require animals to be out on control-grazed pasture. In fact, an outdoor dirt yard is much more unsanitary than a properly bedded structure. For animals the outdoors is not inherently hygienic, any more than being indoors is necessarily unsanitary. Both require careful management, and both can be equally rewarding for man and beast.

To maintain a healthy horde of good bugs in the livestock housing facility requires the proper carbon:nitrogen (C:N) ratio in the bedding. Carbon absorbs the soluble nutrients from excrement and bonds them molecularly in a chemical sponge. If the manure is properly handled it will never smell. Go back and read that last sentence again.

A properly managed livestock housing facility will not smell, period. If just this one principle were adhered to on the farms around the world, agriculture would have a completely different reputation. There is no excuse for smelly livestock facilities. The smell is ammonia (nitrogen) vaporizing. If you smell ammonia, envision dollar signs going off in the air. To maintain a decomposing medium requires a C:N ratio of about 30:1. That is the ideal combination for making good compost.

Just to acquaint you with some of the common ratios, here is a small list (numbers will vary in different books depending on source material for cited research, but they are all in the same ballpark):

Poultry manure is 7:1
Cow manure is 18:1
Good hay is 30:1 — that's why you can just add moisture and watch it compost
Leaves are 40:1
Straw is 100:1
Wood chips are 200:1
Sawdust is 500:1.

The rule of thumb is that if we walk into a livestock housing facility and smell ammonia, we need to add carbon. The deep bedding, or deep litter concept, is not new. The benefits have been documented for decades.

Benefits of Deep Bedding

Holds soluble nutrients in stable medium. One of the biggest economic drains of any farm is loss of nutrients to vaporization and leaching, especially during the winter when the soil is dormant and cannot metabolize them. Holding these nutrients until they can be aerobically composted pays big fertility dividends.

Provides sanitary lounge area for stock. It stops or significantly reduces mastitis in lactating animals, coccidiosis in chickens, scours in baby calves — these are only a few of the benefits documented through deep bedding at proper C:N ratios. Molds and fungi in the medium provide natural antibiotics to the lounging animals. Dairy calves especially respond to clean deep bedding. It is important, however, not to disturb the bedding while the animals are on it. Always wait until no livestock is on the bedding before you turn it or clean it out.

Photo 26-1. *Hay on the left, chips on the right, cleanly bedded cattle eating through a feeder gate: all the ingredients of healthy, comfortable livestock.*

Warm, comfortable lounge area. Because of the heat produced by the decomposition processes in the medium, it never freezes. When it's zero outside and the ground is frozen, animals have an above-freezing area on which to lounge. This has the added benefit of reducing hay or feed requirements because it reduces calories required to maintain body temperature.

Dry lounge area. The insulation of an animal is in the dryness of its coat. A cow with manure or mud caked on its coat is easy prey to sickness. Animal stress is the big killer. The hair, hide or feathers on an animal serve the same function as a jacket on a person. A dry jacket insulates well, but as soon as it gets soaking wet, the insulative properties are gone. A fresh layer of bedding every couple of days provides a dry lounge area, which in turn ensures dry coats for the animals.

Grows bugs. Research done in the 1950s indicated that if at least 12 inches of good C:N bedding were maintained under a housed flock of egg layers, they would pick up all their animal protein from the bugs in the medium. In the Raken house the chickens constantly find bugs in the deep litter. The chickens aerate it and it composts right on the floor. Earthworms come up into the deep bedding in the hoop houses, where our pastured layers live in the coldest winter months.

We move a small "Porkovator" across the bedding to let the pigs dig down deep and bring up the worms. This is a small portable pigpen containing small pigs and we move it twice a day to keep the bedding from becoming capped and use a bigger "machine" to more fully stir the bedding.

In our broiler brooder house, mortality decreases as the bedding builds throughout the season. Between batches, we just stir the bedding with a garden hoe to inject oxygen, add some fresh wood shavings, and put in the new chicks. The deeper the bedding, the healthier the birds.

Although it does require handling carbonaceous material, deep bedding pays for itself in feed savings, animal comfort and performance, and especially in animal health. Animal housing need not be filthy. In fact, it can be much more sanitary than a dirt loafing yard outside. With careful management, it can be a fun place to go and be with your animals.

One of the most important points of this discussion is that sanitation has nothing to do with size. I've seen atrocious parasite loads in tiny homestead situations where animals do not get fresh ground or where chickens are in a little dirt pen. These conditions are no better than a filthy cattle feedlot or factory confinement house. The only difference is magnitude.

Even a confinement turkey house could be a sanitary thing if the population density were cut by about 65 or 70 percent, lots of skylights were added to the roof, 500 percent more carbon were added to the bedding, walls were opened enough to get strong natural ventilation through and some pigs were added to keep the bedding stirred. Of course, the poultry industry considers such a model completely uneconomical and too laborious. It's much cheaper to pump the birds full of pharmaceuticals, hire serfs and blame wildlife for spreading pathogens.

I believe sanitation is one of the most critical areas of plant and animal husbandry neglected by new farmers, primarily because they build conventional structures or overextend themselves with too much too fast. Take care of what you have well before adding anything else.

Sanitation is really not difficult to achieve. For emotional reasons if not for economic ones, it is the first line of offense to making your farm a success. Nothing is as frustrating as sickness, and nothing as exhilarating as vigor. Put attention on sanitation in all your production models and you'll be surprised how many other things will fall into place.

Soil Fertility

In any part of the world, the most fertile areas are generally those undisturbed by man. The hardwood forests of Appalachia once covered soil that went down a foot in black, rich material. Along most edges of forestland where forests have been maintained for a few decades the soil is generally several inches deep in this same rich soil. Just a few feet away, in open land like pasture or cropland, the black runs out in less than an inch, if it exists at all. Centuries of prairie built literally feet of black, humus-rich topsoil. In many areas row crop farmers are desperately trying to hang onto what remains of it.

Truly Sir Albert Howard said it perfectly: it has always been mankind's temptation to turn humus into cash, to capitalize on what nature took centuries to accumulate. Today we've gone beyond humus capitalization to view farming as input and output, much as a factory.

It's time for a paradigm shift in thinking. Rather than farming with inputs and outputs, we need to view our farm as a giant reservoir, collecting solar energy. We want as slow a leak in the dam as possible, all the while letting this huge mass of energy build and build to heights of fertility unimaginable.

I've often looked at especially fertile spots on the farm and daydreamed about what it would be like to have the whole place look like that. One spot, near a stack of concrete blocks by the equipment shed, would probably produce 1,000 cow-days per acre per year worth of grass. Although for certain reasons it may be impossible (soil depth, slope, aspect, etc.) to totally duplicate this performance farm-wide, such dreams are the things that encourage and challenge us.

If I go to that area and do a little digging, the soil has a certain smell, a certain feel. It is full of tiny critters like worms, beetles and centipedes. Agronomists have all sorts of fancy names for soils depending on their parent material. I know that soil testing laboratories establish cation exchange capacity, which measures soil potential.

But I've found that regardless of these scientific measurements and designations, fertility principles span all climates and soil types. In fact, if you don't let the scientists limit you too much, you can totally change the productivity of poor soil to nearly that of your favorite fertile soil spot — given enough time, the right materials and an indomitable spirit.

Much of our farm 35 years ago resembled those Natural Resources Conservation Service (the old Soil Conservation Service) pictures in booklets describing the tragedy of American soil erosion. Many gullies measured 10 feet deep and one measured 14 feet. Most of the land had no black topsoil. In places the red clay and shale came right to the surface. We still have trouble putting in temporary electric fence stakes in some places because the soil is not deep enough to hold them up. Large shale dishes dotted the fields.

In the early days Dad poured concrete in old tires and squished a piece of half-inch pipe into the mushy concrete. When these hardened, we would put them out to hold up electric fence stakes where not enough soil existed. These thin spots have now grassed over enough that we no longer need to use the old concrete-filled tires.

On one of these hillsides about 20 years ago, we were spreading some barn manure on a shaley part when all of a sudden the manure spreader advance chain broke. Nothing to do but shovel it

Photo 27-1. *Animal manure is black gold, and should be revered as such.*

out. Of course, we flung it as far as we could, but right in that spot we put on a liberal application. That 100 square-foot spot has never received any special treatment since. Still, it grows five or six times the forage of the area adjacent to it, greens up earlier in the spring, stays green later in the fall and basically doesn't resemble the surrounding ground at all. The soil is black and loamy, with abundant soil critters. It takes drought better and grows back much faster when mowed or grazed.

Another interesting thing about this spot is that whenever we graze that two- or three-acre shaley paddock, that 100 square foot area gets eaten to the ground in the first 10 minutes, indicating increased palatability. Furthermore, the cattle tend to lounge on that spot, even though it is not on the top of the hill or in the same plane as their normal lounge spots on that field.

Now, if only the rest of that area could be made just like that spot. That is my dream. The exciting part about that dream is that I know the rest of the area can be made like that spot and I know exactly what it would take to get it there — several broken manure spreaders!

We have another field that faces east and was always poor. One year in the early 1960s we had a couple loads of rain spoiled hay that we mixed with some manure to build two compost piles, one on either side of the hay wagon. We fully intended to spread them the following spring, but time got away and we never did spread it. Here again, many many years later, I can show you those spots, the cows can show you those spots, and the earthworms can show you those spots. They are drastically different than the surrounding area.

Soil fertility is not some elusive, nebulous concept that we can never reach. I guarantee that whether your soil is sandy, rocky, loamy or shaley, if you had done the same thing on two spots that we did, the response would be equally dramatic, and equally long lasting. True fertility does not come and go with the season. It can't be bought in a bag.

True fertility comes about as many different factors interact to boost soil organic matter and biological activity. Soil scientists often describe the positive productivity of new ground. The pioneers certainly benefited from new ground. Row crops following forages always produce better than when they follow other row crops or each other. Crop specialists don't have answers to this, except to label it the principle of new ground.

Whole cultures have risen and fallen because they could not sustain the good spots. They couldn't figure out how to make all the area like the good spot, and instead made all the good spots like the bad ones. Organic matter is the key.

Chemical fertilizers require a larger and larger dose each year to get the same kick, much like a drug habit. Based on petroleum, they are acidulated in order to increase solubility. The chemical agriculture mindset is that plants are merely wicks poked into inert material (soil) and should therefore be fed like a sick person on IV tubes. It's essentially a glorified hydroponics system. The acid salts kill the living parts of the soil. The accumulation of these salts can be seen in large "hot spots" where nothing will grow in California's vegetable-producing region. Irrigation speeds up the process as well. Because the salts and chemicals burn the living critters in the soil, they cannibalize in order to survive. This is why we talk about a heavily fertilized farm with dead soil as being "burned out." Too much of anything can cause this. For example, a heavy dose of pure chicken manure, especially in a dry period, will burn plants. We recognize this tendency in manures by calling them "cold" (cow, at C:N ratio of 18:1) and "hot" (at 7:1), and adjust application rates accordingly. Acidulated chemical salt fertilizers are essentially all "hot" but with the added detriment of low residual access. This means that the fertilizer value must be used by plants within roughly 90 days or it becomes inaccessible. Some research indicates that the unused residual actually hampers the efficacy of subsequent fertilizer applications.

When the living parts of the soil decompose, they give off carbon dioxide. When this gas contacts moisture, an acid is formed. In a chemistry lab, if I want to break out the minerals in a rock, I can use many acid reagents. I can use sulfuric, hydrochloric or any number of acids, but the most efficient one is carbonic acid. That is the acid formed when carbon dioxide meets water. It's the one formed in the soil as organic matter decomposes into humus. Carbonic acid is the key to accessing minerals.

Just as each positive has its negative, each positive acid molecule in the soil has an improper synthetic counterpart. In his wonderful book *An Acres U.S.A. Primer,* Charles Walters Jr. points out that urea, the heavily nitrogenous portion of urine, contains one more atomic weight (a neutron) than synthetically manufactured urea. Why is that important? No one knows for sure, but I have a sneaking suspicion the natural number is there for a reason. Oriental custom has long held to the mystery of life, and I believe some of these things fall into that arena.

Whatever you put on the soil that burns your skin, or is otherwise hazardous to you, will also burn up earthworms and soil microbes. More organisms live in one handful of healthy soil than there are people on the face of the Earth. How dare we treat the soil like dirt?

Certainly some good natural soil amendments exist. Certainly you can buy them and probably speed up the healing process. We use an organic, highly mineralized amendment in our compost, and of course put our wood ashes through the composting cycle. When biology acts on minerals, as in a compost pile, it synergizes the minerals so that you can get more benefit from a given poundage. I have heard people who study these things use figures as high as a four-fold increase. In other words, 500 pounds of minerals put through the composting process and applied to the soil is equivalent to 2,000 pounds spread raw. When I say minerals, I am not talking about limestone. I am using the term generically. In composting, one of the key minerals is phosphorus.

I am not a soil scientist, but would suggest that you do some reading on the subject to gain at least some appreciation for just how alive the soil is. If you can just stop and think about how soil builds in nature, you will realize that it likes organic matter, and it likes to be fed on top, not underneath. As my good friend and a truly superb soil scientist Bill Wolf says: "Earthworms want their food on top of their heads."

One pound of organic matter holds four pounds of water. Not only does it allow water to penetrate faster, but the total holding capacity is greatly increased when organic matter is higher. This has everything to do with floods, erosion, drought tolerance and capillary action, as well as aeration, root penetration and cation exchange capacity.

Weeds describe soil fertility. In our area, broom sedge, wild strawberries, blackberry vines and hawkweed all indicate calcium deficiencies. The typical solution is to spread burnt lime. Many times this complicates the problem because magnesium ties up calcium, and most lime available is high-magnesium lime because calcitic lime is light and hard to handle. It tends to spread out instead of staying in a nice pile. Dolomitic lime is high in magnesium and that can be the best material if magnesium is low. Burnt or hydrated lime is treated for quick release and is not beneficial over the long haul.

Composted manure, or just raw manure mixed with bedding, will knock out these weeds and bring on grass and clover as fast as anything. But manure is acidic. When organic matter is where it should be the soil adjusts itself, the carbonic acid breaks out the minerals necessary for proper balances, and true fertility occurs.

Lambsquarters and pigweed indicate high fertility soils; so does cocklebur. In a hardwood forest, spicebushes indicate fertility while mountain laurel indicates poverty. Dandelions fix calcium in the soil. Wild carrot (or Queen Anne's lace, as it is commonly called), and blue weed (chicory) send down deep taproots and translocate soil minerals from the subsoil to the topsoil. In the garden, the best Irish potatoes grow next to a towering weed.

Twenty years ago our farm was covered with dewberry vines and wild strawberries. We've not spread any limestone, and yet you can't find a vine on the place now. Patches of dense broomsedge, indicators of calcium deficiency, are now being replaced by vigorous, healthy red clover plants. Clover indicates high calcium. And yet these two plants are growing side by side. I believe these areas are shifting to higher fertility and the vegetation is only responding to what is righting itself below.

Livestock fed forage from poverty soils will still excrete dung rich in nutrients that differs only slightly from dung excreted from animals fed forages from an extremely fertile soil. The enzymes and body chemistry of living organisms can seemingly create necessary elements out of nothing. Students of such things call this process transmutation. Most scientists don't believe this occurs. A biochemist friend of mine says that the cell mitochondria act as a cyclotron, spinning atomic particles into different molecules.

I don't want to quibble over terminology. All I know is that you can start with poverty soil, and without adding anything but careful management and nutrient cycling, create a soil that does not resemble in appearance, structure or productivity the original poverty material. Ed Faulkner, in *A Second Look,* documents this well.

Wood wastes are, I believe, a poorly tapped resource in speeding up this building process. We piled some wood chips on some grass and removed it a year later. Today, that spot still grows unbelievably lush grass and has earthworm castings two inches high. Similarly, you can spread leaves on an area and see dramatic differences within a year.

Generally, those involved in the input side of agriculture have successfully eliminated nutrient cycling from the minds of those on the output side (producers). Let's look at what nutrients are available on most farms.

First, the biomass created in grasses, shrubs and trees is stored solar energy. A green plant is made up of about 95 percent sunshine (which chlorophyll converts into sugars by photosynthesis) and 5 percent soil elements. In other words, 100 pounds of hay contain 95 pounds of sunshine and 5 pounds of soil elements.

If we bale some 2-ton per acre hay, we take off 3,800 pounds of sun and only 200 pounds of soil material. Certainly roots add some soil humus through decomposition, but not all 200 element-pounds.

Next, realize that a beef cow produces more than 20,000 pounds of manure and urine per year. For conservative figures, let's say she produces 50 pounds per day. Of that 50 pounds, 1.5 pounds of nitrogen, .85 pound is phosphorus and .5 pound is potassium. Feeding the cow half a bale a day, or 20 pounds of hay, she produces 50 pounds out her business end. How's that for return on investment? Those two tons of hay, plus the water the cow drinks, produce 10,000 pounds of manure and urine.

Certainly each farmer has his own special leaky dam problems. To plug all the holes would require shifting production from centralized megacenters to regional food production. It would mean consumers eating a more seasonal diet. It would certainly mean dismantling manure burning utility plants. But don't let the enormity of a pure system keep you from implementing good models in your situation. We need to focus our money and time on the nutrient resources at our disposal, regardless of how big or small they may be.

Back to our story. We've extracted 200 pounds of soil elements (humus, minerals, enzymes) and we return dead roots and 5 tons of manure, which contain 300 pounds of N, 170 pounds of P and 100 pounds of K. Does all this sound too good to be true? Obviously, these figures suggest that buying fertilizer is practically unnecessary.

Before I go too far out on a limb here, let me admit that 2 ton per acre hay is extremely good hay. We assumed that the cows ate all of it and that we lost no nutrients. In the real world, none of this is possible. The cows will not eat every morsel and we cannot capture every single ounce of nutrients. But we can capture most of it *if* we manage it with carbon, composting and correctly timed application.

True, we can buy in organic soil amendments and build up the soil dramatically and rapidly, but that may well destroy our cash flow. It may also push production higher than our ability to manage it. Going slower keeps our cash requirements lower and allows our experience and stewardship level to come up on par with fertility, in a balanced approach.

The other idea I want to make clear here is the economic sense in buying food to put through livestock as a fertilizer technique. For example, we buy tons and tons of chicken feed for our birds each year. This is a positive nutrient flow into the farm, just like buying fertilizer. But unlike fertilizer, we net $1,200-$1,500 per acre to apply this fertilizer. Running wholesale inputs through animals and building fertility with the manures is generally more profitable than simply buying the fertilizer. The animals, especially if they are on pasture, have a much more balanced manure nutrient profile than animals in confinement factory concentration camps. The nutrient ratios are completely different. Beware of buying chicken litter from concentration camps.

Some farmers who raise herbivores do not make hay, but rather buy all their hay, and thus import nutrients that way. I realize that in the pure sense of the word, this may not be the most sustainable system, but let's not go bankrupt trying to be sustainable. Besides, as I've said before, mutual dependency is a way to build community. Two farmers with different complementary centerpiece enterprises may collaborate for mutual benefit. Both neighbors export and import nutrients, and that is one way to build community while enabling each to make a living.

Some people ask me why we don't grow our own grain. That is simple. This farm was part of the breadbasket of the Confederacy for a century, producing grain on hillsides you can hardly stand on. The soil is gone and very few acres exist that would come close to producing a crop of grain naturally. Besides, growing a little grain is like being a little pregnant. You either are or you aren't.

In today's economy, I don't believe it's wise to produce small acreages of grain unless you're already tooled up for it. But we have neighbors that do have some good ground and already own the equipment so we let them produce the grain. The big thing to remember is that if the 70 percent of all grain grown in this country to feed multi-stomached animals were eliminated, we wouldn't be producing grain on marginal lands and even small farms could produce it economically because the value would be so high. Once farmers discover the value of grass, the price of grain will need to double in order to entice some farmers to grow it. When that happens, we'll see more local production for local use, smaller fields and smaller equipment.

This is not a book about saving the world. This is about making a living at profitable agriculture. Using grain from the neighbors comes nowhere close to violating my worldview principles. Tapping into the "cheap food" policy by bringing in cheap grain for your pastured poultry, or hay for the cows, as a means of being paid for building fertility is a wise policy.

Now back to our cows returning manure from the hay they

ate. In addition to the manure, we have earthworms to literally create fertile soil. A healthy population of 12 earthworms per square foot will produce 6 inches of highly mineralized soil castings per year. Rich in humic acids, these castings contain 5 times the nitrogen, 1.5 times the calcium, 7 times the phosphorus and 11 times the potash (to name just a few) as the original soil. Castings are much richer in nutrients than manure, and at 3 inches per year add 50 tons per acre. Those worms increase aeration and tilth, producing castings, all while we sleep.

Soil does not deepen, it "uppens," as Andy Lee says. In Daniel's rabbit pastures, where we put down poultry netting to keep the rabbits from digging out of the "Harepens" we have a good reference point to measure this. When we began this technique, we figured we would need to replace the wire every 5 years because it would rust out. Instead, we are having to replace the wire every 5 years because the soil builds up over it enough to let the rabbits dig out between the new soil surface and the now buried wire. The soil "uppens" that fast under intensive use. Of course, Daniel imports rabbit feed in this process, but the rabbits voluntarily cut their intake of purchased feed by 60 percent. It's a dramatic step in the right direction.

One of my dreams is that we will have to replace all the fenceposts on the farm because the soil will just build up over all of them. Wouldn't that be exciting?

This land was abused for decades and decades, and to take remedial action like buying in grain or hauling in wood chips and leaves from the city dump, or using some petroleum to chip our own material, is the least we can do to stimulate the healing. Diverting grain here does not have a spit in the ocean's effect on the grain cartel or the erosion factor. Diverting city leaves here may take some diesel fuel, but it keeps them out of the county landfill. Yes, it would be wonderful if everyone in the city would compost their leaves and turn their lawns into gardens. But we must crawl before we walk, and right now we have acres and acres of wounded soil in the coun-

Photo 27-2. *Carbon is the centerpiece of soil fertility. Acquire as much of it as you can. This is a pile of pine bark mulch from a fence post manufacturer.*

tryside that can heal from desperately needed organic material and nutrient inputs.

Woody species not fit for livestock consumption can be chipped and added to the soil. Lapwood (the treetops) left from timber thinnings or harvests, "weed trees" removed in land clearing or upgrading and any other woody material can be chipped. Perhaps you could purchase a nice chipper and offer this service to neighbors and city people. They pay you $40 per hour to chip their material and you haul it back to your farm.

Sawdust can also be used. Trees can be viewed as giant solar collectors storing energy as carbon. The same energy that heats a house on a snowy evening can grow a crop of hay or potatoes. Here is energy (fertilizer) that a farmer can grow on his own place or acquire from nearby sources. I met a dairy farmer in Ontario who grows 2 acres of long-stemmed rye each year just for carbon.

With all this in mind, why do we need to buy fertilizer? It's obvious that at our fingertips are all the nutrients necessary for high fertility. So what's the problem with fertility? Where is the leak?

Let's look at the leaks. First, the biomass production. If we put on acidulated fertilizers that kill microorganisms, root and biomass decomposers, the soil can't build. Small amounts of chemical fertilizer spread over numerous applications reduce the negative effects. Furthermore, erosion, especially on cultivated land, takes a tremendous toll. Leaching of minerals into the subsoil or groundwater occurs in varying degrees, usually in direct relation to the organic matter content. Cultivation and chemicals, from pesticides and herbicides to synthetic fertilizers, kill microorganisms that form the organic matter base. Tillage oxidizes organic matter away, similar to the rusting principle.

Now, remember the 50 pounds of manure and urine produced daily by the cow? In the raw state, dung nutrients are highly soluble.

Photo 27-3. *Bedding down the hayshed with carbon provides the critical link to maintaining soil fertility: minimize nutrient losses and synergize resources through biological activity.*

Solubility generally means that if water hits them, nutrients run off, and if air hits them, they fly away. Agronomists call this leaching and vaporizing. Since most of the nitrogen, in the form of ammonia, is in the urine and the water of the manure, most of it is lost on a sunny day through vaporization. Half the nitrogen in freshly spread raw manure can vaporize in only one day — smell it?

Many of the dung nutrients are lost in confined feeding facilities, watering areas, under shade trees and in pasture campsites where they are not returned to the area whence the feed came. An increasingly serious problem is sterile dung. Some composting consultants are reporting manure so sterile that it will not decompose. Systemic wormers and grubicides kill some 400 species of dung beetles and microorganic decay critters that inhabit soils and decompose dung pats.

Can earthworms make up the shortfall? Realize that these natural tillage critters are the pinnacle of soil organic matter. Anything that hurts humus also hurts earthworms. Tillage, chemicals, low organic matter, low soil moisture, constant livestock impaction and absence of soil mulch all discourage or kill earthworms. After a warm rain (anything over 50 degrees F) take a walk and look for castings. They should nearly cover the soil. Often you may find that there is only one thin mound per square foot. Ideally, the mounds should be an inch high with practically no undisturbed soil between them. Monitoring earthworm castings beats any soil test man can invent for determining fertility.

The energy potential of woody species and trees is simply not being used. Shrubs often get sprayed in annual eradication efforts. Cattlemen too often view forest acreage as a barricade in the way of the mowing machine. Trees, which are four times more efficient converters of sunlight to energy than grasses, are generally viewed as unnecessary and non-complementary to cattle, or open land production. Worse, foresters often burn residues and send all that energy up in smoke.

In our county, tree-pruning experts dump most of their chips at the landfill. I asked the operator there why they had all those chips piled to the side. He said that they mix them in the top foot of capping soil, when they finish off an area, and it makes the soil so black they don't need to buy fertilizer. Why can't farmers see what the landfill operators see?

To wrap up this discussion, here are some recommendations for building fertility:

1. *Feed the soil only things that will stimulate organic activity.* Compost is the best, but any organic matter will do.

2. *Use carbon-rich deep bedding to minimize dormant season vaporization or leaching.* We feed hay in the barn through a feeder gate that can be raised and lowered four feet to accommodate bedding buildup. Wood chips give us the carbon for the bedding base, and the carbon bonds molecularly to the nutrients like a chemical sponge. We compost this material, which reduces solubility by about 40 percent, and return it to the hay fields after the spring flush of grass growth. This feeds the soil when it can best uptake it.

3. *Use controlled grazing for all animals.* If you only have two goats, that may mean a 10 foot x 10 foot portable corral, but that will keep them under control, give the grass much needed rest, and keep the animals away from parasites. This cycles the manure onto the field instead of concentrating it on campsites.

4. *Protect water supplies.* Fence water facilities and use portable troughs that can be moved around with stock to minimize impaction in watering areas.

5. *Reduce tillage.* The less you till, the more undisturbed biological activity can occur. This helps keep the ground covered with mulch or sod, additionally stimulating microorganic activity.

6. *Build ponds and use them for strategic irrigation.* Build berms or ditches to control the flow of water across the landscape by slowing it down to give insoak time.
7. *Haul in organic matter like your life depends on it.* We pay a utility line chipper crew $10 per load to dump here at the farm. Clean out horse stables for the material.

8. *Mulch the garden.* Although mulch can be considered labor intensive, it accomplishes multiple things in one task: water retention, fertilizer and weed control.

9. *Apply raw minerals only,* and as much as you can reasonably afford.

10. *Use small animals to build fertility fastest.* The smaller the animal, the more efficiently it will build fertility per square foot, pound for pound of live weight. However, the smaller the animal, the harder it is to feed it completely on your own homegrown material.

Much of our nutrient cycling breaks down because we impose on the land a production scheme it is incapable of recycling. Cropping slopes, grazing steep hillsides, continuous grazing, most feedlots — all of these extract more than they can possibly replace and are inherently degenerative. Although I appreciate climatic, topographic and geographic differences, in every place, with real diligence applied to true fertility and nutrient cycling, farmers can sustain virgin soil productivity. True fertility is within your reach. Go for it.

Seasonality

Getting in sync with the seasons offers advantages in both production and processing. Seasons exist for a reason, and to ignore the cyclical up time and down time militates against nature.

In dairying, for example, nearly 70 percent of the farm expenses are attributable to trying to make cows lactate in the winter. In nature, herbivores do not lactate in the winter. They are in a holding pattern, actually losing weight, waiting for the spring lush. Mature females have weaned off their young and are simply maintaining body weight as the new baby grows. But mature animals, just like mature people, do not need much to eat when they are not working hard. You have heard of middle-aged spread, haven't you?

Farming used to be more seasonal. Farmers would butcher beeves and hogs in the late fall, after the temperature cooled down but before feeding the animals would require stockpiles of stored feed. Chickens naturally lay best in the spring during increasing daylight. The spring egg flush provides excess for hatching broilers in the spring, to be harvested throughout the summer. Egg production drops way off in the fall and early winter. Farmers would always set a few hens to hatch chicks during this spring overage. That meant chicks would hatch right when the greatest number of grass-

Photo 28-1. *Calve when the deer are fawning: one of the cardinal rules in profitable beef production.*

hoppers and crickets were available. Turkey poults require exceptionally high levels of bugs. As they grow, their protein requirements diminish, right along with diminishing bug levels late in the season. Their starch requirements increase, and late in the season is when grasses and weeds (including domestic varieties like small grain and corn) make seedheads, which are higher in starch and lower in protein. As fall and Thanksgiving approach, their nutrient requirements shift to exactly what nature provides at that time. Right before winter, turkeys are perfect for harvest: it is the "fullness of time."

Beef and pork are heating meats; chicken is a cooling meat. Isn't it perfect to eat beef and pork in the winter and poultry in the summer? Beef and pork are the most succulent in the late fall.

Vegetables and fruits certainly go through seasonal shifts. In Virginia, head lettuce, onions and radishes can be planted in the early spring, weeks before the warmer weather plants like corn, green beans and squash. Some grow best when it is a little dry and others like wet feet. Anyone who has gardened for decades knows that some years are more conducive to potatoes and other years are better for cucumbers.

As soon as we try to buck the season, production costs escalate. Generally off-season vegetable and fruit production may compensate for this extra cost but meat, poultry, dairy and eggs do not. I think the reason is because vegetables have such a quick cash turn-around time. The cost of production is easier to figure because it is compressed into that single crop.

Livestock, on the other hand, has a long cash turn-around time. Additionally, people are still used to seasonal vegetable production. We have canning and freezing season, and even the average consumer is somewhat aware of that cycle. But intensive environmentally controlled factory livestock production obscures similar cycles for meat and dairy.

Only you can determine if the extra effort and cost of producing tomatoes two months before the normal season can be recovered with a higher price. Greenhouses, fuel and labor, not to mention additional risk, all require that off-season vegetable production be priced at least the going rate during the season. My point here is not to deter you completely from producing things out of season, but rather to warn you about the financial pitfalls associated with such a policy.

The average consumer is completely out of touch with natural seasons. We go to the supermarket and expect an array of fresh meats, vegetables and fruits all year long. We expect to walk in when the temperature outside is zero and find fresh snap beans, cucumbers, Delmonico steak and broilers. Does it occur to us to think about what it must have taken to put that stuff on the shelf at this time of the year? It just all seems to appear by magic.

It took a tremendous amount of petroleum and pharmaceuticals, not to mention a lot of disagreeable, cold-finger farmers' mornings, to get it all there this time of year.

Stretching the season without overusing energy and labor is a commendable goal, but only if the price increase merits such expenditure. Probably Eliot Coleman has done as much as anyone to develop efficient models in this regard. His *Four Seasons' Harvest is* a classic already. Double cold frames, double hoophouses and the like can capture solar energy to give you a production edge on the competition. But no matter what you use, as you push the seasonal boundaries, the costs escalate dramatically.

I would encourage all beginners to stick with the season, realizing that is the most forgiving, cheapest means of production, and then push the boundaries only after you develop proficiency during the easy periods.

Let's examine some of the common ways farmers lose money by refusing to incorporate seasonality into their production models. I mentioned dairying above, and it is still probably the most glaring example. In nature, herbivores go through a kind of fasting period in the winter. This accomplishes two things:

(1) It purges the system of toxins and gunk. Just as people fast to clean themselves out, a lean period helps animals stay thrifty and healthy. In nature, fat animals are highly dysfunctional; you just do not see fat animals in nature except right at the end of the lush season. Fat cows do not rebreed well; fat rabbits have heart attacks and become infertile.

This lean period actually stimulates health and vitality. Feeding calves high energy feed during the winter, therefore, is not only incredibly costly in terms of harvest, storage, feed handling and manure handling, but it also is not that healthy for the animal. Overly conditioning (condition score is the amount of fat and flesh a cow is carrying on its skeleton) dry beef cows is especially bad because they put their extra energy into the developing fetus. Then they have delivery trouble because the calf is too big.

(2) It facilitates compensatory gain in the spring. If we keep cattle gaining fast during the winter, they will gain no faster and maybe even slow down when the spring flush of grass comes. But if we just maintain them, at a relatively low performance rate during

the winter, so that they just gain in frame but not in flesh, then when they hit spring grass they will compensate for the winter's lean and flesh out rapidly. Within 100 days you cannot tell the difference between a calf that gained 1.75 pounds a day during the winter and one that gained .5 pounds; in the spring, the one will drop to 1.5 pounds daily gain and the other will go as high as 4 pounds.

Putting on gain during the winter is costly and unhealthy. I explain this principle in much more detail in *Salad Bar Beef*. The same general principle exists for all types of livestock. Of course this seasonal production offers a marketing challenge, but right now wholesale livestock markets do not compensate for off-season production.

Broilers cost no more in the winter when houses must be heated than they do in the summer. Actually, factory farming completely nullifies seasonal advantages. The summer advantage for broiler production, for example, is offset by the high ventilation bill.

Just a word about forestry on this point. We reserve our work in the woods for fall and winter. One reason is snakes. Another reason is bees. More importantly, where the sap is affects wood quality. Wood cut when the sap is up is essentially unusable for veneer. Low sap encourages twist-free drying and rot resistance. Especially in some of the rot-resistant deciduous trees like Osage orange and locust, you want to harvest when the sap is down. In fact, old-timers go so far as to say the sap should be down and the moon should be waning if you want a post to last.

Firewood cut in the summer never seems to dry out and burn as well as wood cut in the winter. We like to stack all of our branches when we work in the woods. That keeps the area tidier and provides a home to rabbits and chipmunks for several years as the pile rots. Green leaves nearly double the weight of the branches, making piling much more difficult. It's even hard to swing a branch through the air because the leaves create air resistance.

Winter is also our slow time with the livestock and garden. If we run the farm like a factory, without letup and at full throttle, our life deteriorates into a sameness that is emotionally devastating. One of the complaints about farming is the drudgery. It's the same thing day after day. But as we get in touch with the seasons, every day is new. About the time we get tired of one thing, it's time to start something new. This keeps a freshness in your mind and a spring in your step because you know that this job will give way to a different one.

Nothing is more emotionally debilitating than a factory confinement livestock facility. Cranking through animals is the name of the game. Anything that is factory-oriented, whether it is hydroponics or greenhouse production, can be demoralizing if there is not a break. As a farmer, you probably will not have the luxury of hiring labor for some time, and tying yourself down to a factory model will burn you out quicker than anything will.

As your success improves, additional help, partners and competent friends can give you some free time, perhaps. But as a general rule, plan to be **IT**.

Diversifying enterprises is of great help. Except for harvest, orchard and vineyard type enterprises that require pruning need more dormant season time. Peak vegetable time requirements, of course, are in the summer. Livestock is lower in winter and higher in summer. Tucked around these peak times you can schedule your agritourism, agrirecreation and agrieducation events. Try not to have multiple enterprises peaking at the same time.

For example, keep your livestock birthing times separate. If you have both sheep and cows, plan to stagger the birthing time by a month. If you're running pastured poultry, you may want to concentrate on late season vegetables, where the bulk of harvest would be after first frost and after the last batch of chickens. Something like potatoes, winter squash and a Thanksgiving festival/hayride for agrirecreation would mesh nicely.

Just to give you a rundown of how we've developed this idea on our own farm, let me go through the year with a brief description of what is going on so you can see how the workload changes throughout the seasons.

JANUARY Nothing is on pasture, so no fences or pens to move. Daily chores take less than an hour. We spend large portions of the day reading and relaxing. We order seeds for spring and try to plant some tomatoes, peppers and summer squash in flats to be set into the hoophouses the first of March. On not-too-cold days we'll cut firewood to put on next year's seasoned pile.

FEBRUARY Still nothing is on pasture and chores are less than an hour a day. We're still hibernating, reading and relaxing. We continue to enjoy the fireplace, cook some marshmallows, take short trips to see friends, and begin planning for spring. Toward the end of the month we'll usually have a warm spell in which we might set some fenceposts or do some construction projects and prune grapevines. Anything we want to cut down for seasoned firewood for the following winter, we try to have down by the end of the month. Target date for finishing all woods work: end of February.

MARCH Things start to speed up. The first week, we take layers out of the hoophouse and get them on pasture. We do fence repair and get brooder houses ready. Cow's come out on pasture and begin grazing; pigs go into hay shed to begin pigaerating compost. The first batch of chicks arrives around the 20th. As soon as hoophouses are empty, we plant vegetables, both seeds and sets. By the end of the month, rabbits come out of hoophouses and back onto pasture. The spring newsletter goes out.

APRIL Orders from the newsletter begin pouring in and we know the season is upon us. Customers write notes of encouragement and we can now see income for the rest of the year. Some early vegetables can go into the garden, we work on construction projects like buildings or maintenance and toward the end of the month get the first batch of broilers out onto pasture. This is very much the big transition month from winter to summer. Cows start calving and grass begins to take off by the end of the month. Farmers' Market starts each Saturday morning and we feel like we're back into circulation.

MAY More broilers come. By now we're up to three batches, three weeks apart. Calving continues and egg production peaks for the year. We process the first batch of broilers after the middle of the month and that brings in our first big paychecks of the season. Pigs finish composting the cattle bedding and we take them up to the pig pasture. Equipment must be readied for haymaking. We plant summer vegetables and try to basically finish getting the garden in.

JUNE It's haymaking time, the most stressful two weeks of the year. We watch the weather, pray that the machinery holds together, and try to finish in between the first and second broiler batch processing. Daily chores now take three people three hours a day and we're sprinting. The cows have calved and pasture rotation settled into an easy rhythm. We spread hayshed compost on hay ground and try to do just enough maintenance on the garden that it doesn't get completely away from us.

JULY Things slow down from haymaking. Now we're in the middle of summer. Hot days are not conducive to strenuous work, so we do some repair and maintenance jobs. We normally do some bandsaw milling of our logs into boards and have a few cookouts. Daily chores are still running three hours but not as much is pushing us. Generally we harvest a batch of pigs and start some new ones.

AUGUST Garden vegetables begin really coming in. This is canning and freezing month. Thanksgiving turkey poults arrive when the last batch of broilers goes out onto pasture. Things look fairly dry and we're beginning to really watch the grass growth to see how much forage we have ahead of the cows. We continue on lumbering and construction projects, but take it easy many afternoons.

SEPTEMBER The season is winding down. Turkey poults are on pasture. Brooder houses stand empty. We continue to harvest garden bounty and prepare for dressing beef and pork. We haul out a couple loads of firewood for the early bird customers. Fish begin biting at the ponds.

OCTOBER We dress the last broilers and move turkeys into some of the vacated pens. Many pens stand empty. The garden wanes but grapes are ready to pick and juice, as are elderberries. Sweet potatoes, Irish potatoes, pumpkins and winter squash get stored in the cellar. We finish all the early firewood orders and take beef and pork to slaughter, cutting our livestock numbers drastically.

NOVEMBER Early in the month we begin putting layers and rabbits back in the hoophouses. We wean calves, haul out chicken litter from brooder houses and ready the hay shed for winter. At the end of the month all the turkeys are dressed for Thanksgiving, which vacates all the field pens. The only animals still in the field are cattle. Chore time drops to one hour a day.

DECEMBER It's time to rest, spend money before the end of the year and finish grazing most of the pasture. We spend almost no time outside. We enjoy visiting friends and taking time off.

Obviously, the seasons provide for both hectic and relaxing times; for your spiritual, emotional and physical well-being, you need this diversity. Rather than bucking the season, enjoy it and capitalize on the opportunities change affords.

Synergism, Stacking and Complementary Enterprises

Increasing income per unit of land, or per foot of building space, is not *a* priority, but *the* priority of the small farm or homestead.

Modern farmers, bowing to the paradigm and so-called efficiency of industrialization, have gone belly-up streamlining operations into single-purpose production models. Confinement swine houses, poultry houses, dairy barns and feedlots produce only one item.

We farmers often refer to ourselves as "dairymen" or "beef cattlemen" or "hog farmers." Ask most farmers what they produce, and they will respond with a single item: apples, blackberries, hogs, eggs, milk, corn, soybeans, etc.

In nature, you never see single-use acreage. Every acre produces low-growing vines, tree-climbing vines, low-growing flowers and herbs, higher-growing shrubs, medium-height trees and tall trees. A prairie grows generally at least 100 varieties of plants on any given acre. Some are shade tolerant while others need full light; some grow vertical and others grow more horizontal; some like it hot and others like it cooler; some like it dry and some prefer moist conditions.

A pond or lake, likewise, contains a multitude of different species, occupying different strata of the water. Along the edges are hydrophilic plants and further into the water aquatic plants take over. Some animals live in the top foot of water while others, like carp, stay on the bottom in the cooler water, eating things that drop to the pond floor.

But in our blind, headlong rush to embrace industrial, factory-type models in the biological world of agriculture, we have spurned the lessons of nature and instead adopted single-use models. If you think I am stretching the issue, how about asking a beef farmer what he thinks of running free-range laying chickens behind the cattle to de-fly the pasture and spread cow paddies? He'll say: "Hmmmmmph. I'm a cattleman, not a chicken farmer." End of conversation. And so land that used to produce a myriad of species, including wolves and turkeys, is reduced to a single species — cattle.

Orchardists are in the fruit business. How many would consider running chickens through their orchards to clean up dropped

Photo 29-1. *A portable pen with laying hens moved under orchard trees gives my neighbor and friend, Tom Womack, a synergistic and stacking model: the chickens lay eggs and fertilize the orchard, debug around the trees, and convert both pests and dropped fruit into eggs. Chickens help mow the grass, but also grow more, which can feed a milk cow, sheep or beef animals as well. Multiple crops off the same square footage of land greatly increases income per land dollar invested.*

fruit, debug the sod, and clean around the base of the trees? Better yet, how about adding sheep to mow the grass, fertilize the trees (that's where they go for shade) and then run the chickens behind the sheep to clean up the sheep parasites, debug the dropped fruit, and clean around the base of the trees? If the fruit yields $2,000 per acre, the sheep can easily add another $500 and the chickens another $2,000.

This brings us to the concept of stacking, or tiering, which is exactly the model God used in building nature. Natural patterns of production are cyclical, seasonal, and diversified — modern farming deplores all of these.

Linear reductionist Western thinking dominates agricultural thinking. A poultry house produces poultry. It doesn't matter that the manure menaces the neighborhood and pollutes the groundwater. The feed goes in the front door and the manure goes out the back door; input, output — just like a factory. But if we view all the processes as subsets of other practices, models become a group of circles rather than inflow-outflow single-purpose procedures.

Those of us who have opted out of conventional educational paradigms see this easily in compartmentalization. You and I know that you can't study any subject divorced from all other subjects and get a balanced perspective of the material. We must view education as a whole; nothing is amoral or unrelated.

As farmers, and especially scale small farmers, we must be keen on trying to build circles into the farm production models so that the output of one thing is the input of another; so that the liability of one element is the asset of another. This significantly increases the production per foot or per acre, allowing us to increase sales without increasing capital-intensive land, buildings or machinery.

Let me give you some examples of stacking principles that we use here and others that I've been privileged to see or hear about. Certainly one of the critical factors in any of these models is careful control and planning — management. These are designed systems, using nature as a blueprint, to enhance the productivity of the land.

A citrus operation in Belize runs chickens in pens underneath the trees. The chickens pick up dropped fruit, debug it, fertilize and produce eggs that these folks can sell at double normal prices because they are so good.

A friend in West Virginia breeds and sells guard donkeys for sheep. These miniature donkeys follow the cattle in the winter, thriving on the hay the cows spoiled. Parasites do not cross-speciate, so the donkeys have no problem gleaning what the cows waste.

A farmer in Arizona irrigates a vineyard and runs chicken pens between the rows. The most striking thing about this vineyard is the wide distance between rows. This wide spacing was the breakthrough in getting more use out of the land because it opened up more opportunities for using that space. When the space is only eight feet or less, use is limited. These folks use chickens in portable pens to prepare and fertilize the soil, then plant garlic and vegetables in the space between rows. In one year, then, they produce a crop of grapes, eggs or poultry, and vegetables.

In our hoophouses, where we keep laying hens in the winter, chickens are on the deep bedding floor. On one side are earthworm beds protected by wire cloches. The chickens roost on the wire cloches, dropping their manure onto the earthworm beds. As the earthworms proliferate, we can feed them to the chickens as fresh animal protein during the winter. In the spring, when the chickens go out, we plant vegetables into the composted bedding — composted because the carbon-nitrogen ratio was kept correct and the chickens did the aerating. To reduce capping, pigs in a portable pen can be walked down through the house to further stir the bedding.

One of the biggest problems in greenhouses is bugs. Between plant growing seasons, laying hens are back in the house, completely debugging it. The chickens provide the fertilizer for the tomatoes. Potential annual production from each 2,400 square foot hoop house, which cost $2,500 to erect, is:

$6,000 in eggs (4,000 dozen eggs at $1.50 per dozen)
$4,000 rabbits (60 litters at 8 per, $2.85 per pound)
$2,000 in vegetables, including tomatoes (1,000 lbs. at $1.50 per lb.)
$2,400 in pork (6 big hogs at 200 lbs. carcass, $2 per pound)
$1,000 in compost—fertilizer (80 cubic yards)
$1,000 in earthworms — savings in animal protein or sold as live bait

This is a dramatic representation of what happens with stacking. Often the income can be doubled and tripled, without any additional capital investment, just by combining and tiering the plants and animals.

J. Russell Smith, author of the classic *Tree Crops,* envisioned multi-storied species in pastures. The first level would be grazing animals, then low-growing fruits like apples or peaches, then mid-level trees like mulberry or cherry and then tall growers like nut trees. This stacking captures more sunlight and converts solar energy into dollars far more efficiently than a single-tier like corn or soybeans.

In the pasture, we run broilers, cattle, layers in a portable chicken house we call an Eggmobile, and turkeys. Sheep would be a logical addition. Imagine the return from these enterprises, given in annual return per acre:

$500 beef
$2,500 poultry
$40 eggs
$1,000 turkeys
$500 sheep

As you can see, pasture value suddenly takes on new meaning when we begin adding enterprises that complement rather than compete. Certainly all these critters are not on the same square foot at the same time — they are cycled through before and after each other based on a host of variables that the manager deems appropriate.

Variations on this theme occur all over the world. I've seen, for example, pigs on a slatted floor fed greens produced from a lattice/soil roof shelter fertilized with the pigs' excrement. Instead of just building a roof, a latticework roof with soil and straw mixed in to grow plants and vegetables can enhance the production of a building.

In some areas, caged animals like rabbits can be raised over a pond of catfish or carp, which in turn feed off the rabbit droppings. Placing a Japanese beetle trap over the pond or over a chicken pen is a great way to stack and get free feed.

Our Raken house is a combination Rabbit-Chicken house. When anyone steps into it, the first impression is: "Wow! Look at all the activity! But it's contented activity and it doesn't smell." Breeding stock are at eye level and chickens are on the floor, scratching through what the rabbits drop and aerating the bedding. Although

Photo 29-2. *The "Raken House" illustrates the stacking concept. Son Daniel's breeding rabbits are suspended off the floor and chickens scratch through the droppings, adding a second species and additional product — eggs — to the building.*

Photo 29-3. *Running pigs underneath rabbit pens creates a two-tier production model from a single roof, doubling the income per square foot without the odors and disease that come from single-species factory confinement facilities.*

we have at least as many pounds of animals per cubic foot of interior space than commercial rabbit or chicken houses, we do not have the normal disease problems because the multiple species act in symbiosis.

Pathogens thrive in monocultures and in single-species systems, where these microscopic critters do not have to spend all day (a whole life for some of them) crawling across an unacceptable type of hair, feather or manure drop to get to a proper host. As a result, we can produce as much or more food in the same space as the commercial intensive factory operations, but without the health and smell problems.

As we double up the production, the income potential is mind-boggling. For example, the Raken house, which is a tad larger than a two-car garage, generates $9,000 per year in gross income and about $5,000 net, not counting the compost. I'll put $7 per square foot net up against any commercial industrialized mega-farm in the country.

One important element is to maintain the natural chain of who follows whom. For example, in the hay shed the pigs follow the cows to root through the bedding and make compost, rather than the other way around. If chickens and cows are together, chickens manuring in cow feeders can be a real sanitation problem. But birds naturally follow herbivores.

Permutations on this theme abound. Trellising beans on corn plants is one of the oldest stacking concepts we know about. Master gardeners have developed highly sophisticated symbiotic and stacking models. Double cropping is a classic. Angled fencing panels as a cucumber trellis and growing head lettuce under the shade is a good cool season extender for hot summer months.

Vertical plantings of anything give a greater production per foot. Adding animals to the mix simply increases the possibilities that much more. We bring the Eggmobile up by the garden once a month or so and let the chickens debug the garden. If we leave it more than a day, the chickens begin eating the vegetables. Timing is everything.

Incorporating ponds helps, too. One reason ponds produce more food per surface-acre than land is because of the vertical growth medium in the pond. Everything does not have to stand on the floor, but rather can spread out through the water. Imagine if cows could levitate and eat tall plants 10 feet in the air. That's the idea.

Additionally, the animals in a pond do not need to expend as much energy moving around because they do not have to work against gravity. Every time a cow stands up, she has to work against gravity. Every step she takes, she's working against gravity. But 1,000 pounds of fish can move through the water without working against gravity. That is part of the pond's efficiency.

Using the pond as a microclimate to reduce frost damage on the leeward side due to the heat-sink properties of water allows you to get multiple functions from the pond. One idea we've thought of but have not used is placing a screened cage over the water to receive dead carcasses — chickens, rats, processing wastes, etc. As the flies lay eggs, which hatch into maggots, the maggots fall down into the pond to be eaten by fish. As long as this was not right by your house creating a noxious smell, it would certainly be a way to dispose of animal protein in a synergistic way.

One fellow built a turkey mobile similar to a glorified drying rack. He kept water and grit there, but otherwise the turkeys cleaned up spilled feed from a pastured poultry enterprise. The turkeys essentially ate for free harvesting the waste of the chickens. At night, they roosted on the mobile rack and that protected them from predators.

Grazing animals among Christmas trees is another way to stack. Cattle will not eat the trees as long as you limit their exposure and feed a good mineral. One of the main reasons cattle like young evergreens is that they concentrate phosphorus and most farmers do not feed their cattle adequate minerals. By feeding kelp and moving the cows daily through the tree plantation, you can get mowing and manuring. Instead of being liabilities, these are turned into assets.

This whole concept is best explained and illustrated in *Permaculture: A Designer's Manual* by Bill Mollison, or many of the other permaculture books available. Although practitioners admit that the movement is weak on livestock, these folks have definitely been in the forefront of this whole design and integration process.

I think this is key to maximizing the yield from your acreage. In any farm, the expensive part is the land. Structures, especially if they are built for function instead of form, are relatively inexpensive. The other aspect of this is to look at everything you do as a circle, so that the end of one enterprise forms the beginning of another. The more the production circles intersect the better. We want to fully capture all the waste or byproducts of one enterprise by beginning another enterprise with those elements.

As we tie in all the elements, not only are we constantly creating new enterprises that allow the farm to become more productive and pay for additional warm bodies, we also spread our production base and the whole farm becomes more stable. The more we study these interrelationships, the more we realize that the limiting factor on virtually every farm is not fertilizer, acreage or genetics; it is full utilization of the resource potential. The weak link is our inability to think creatively enough to capitalize on the full range of resources under our feet. The gold mine is there. We just need to tap the potential.

The reason this is so difficult is that thinking creatively, and exposing ourselves to new ideas, takes a lot of work. It is much easier to just buy the latest gadget, the latest gizmo. It's must easier to make excuses: "If I just had that guy's tractor — the one that runs." "If I just had a better education." "If I'd had your parents instead of mine." "If I just had a bull that looked like that." "If I just had the land over there instead of the land over here." "If I just had that manager instead of the one I've got."

We can think up a million excuses, but in the final analysis, I have yet to be on a farm that has exhausted its income-producing and resource-enhancing potential. Let's commit ourselves to looking afresh at our resources, and begin the fascinating journey of putting together the puzzle pieces in a way that the whole is worth more than the sum of the parts.

Chapter 30

Reducing Costs

Whenever I see cost figures for a normal farming operation, I almost go into shock at some of the staggering numbers. True, any enterprise will require expenses, but that doesn't mean we need to see how many zeros we can put on the figures.

Almost anything can be done expensively or inexpensively. Again, the big battle is between your ears. "Keeping up appearances" is more than the name of a British sitcom; it spells doom for many farmers. Do not fall into that trap.

Around our farm we practice function over form. Never the other way around. Simplicity is elegance. When people ask me what I do for a living, I sometimes respond that I spend my time trying to simplify. The easiest thing to do is to complicate things; the hardest thing is to keep them simple.

For example, we took over management of a neighbor's farm a few years ago. It was 120 acres of rolling land, some good and some not so good. We installed 3 miles of permanent electric fence and 6,500 feet of water line in 21 man-days for $2,000. Even counting our labor costs at $100 per man-day, that would only be $4,100 to completely fence and water a 120-acre farm. We used above-ground piping, a portable water trough which we made by cutting the bottom third out of a 55-gallon drum, and used our own locust stakes that we split with wedges in the shop on rainy spring days.

Using your off time, or what would otherwise be unproductive days, to generate resources is a way to get paid for those days. Remember that the dollars you save are worth $1.30-1.50 because you don't have to pay taxes on the income in order to expense it. In other words, if I can keep from buying a $2 electric fence stake and use one that I made on an otherwise do-nothing day, my own stake is worth as much as $3.

Anyway, to stay with the story of water and fencing, a friend of mine about the same time went down for cost-share money from the Natural Resources Conservation Service. They sent out a technician to draw up a pretty map and sketch in the water, using an existing well, and fencing for a 13-acre field. All this fellow had was a 13-acre field.

Of course, they specified an underground water line with companion electric line going to a freeze-proof concrete permanent water tank in the middle of the field servicing four quadrants of multistrand permanent electric fenced paddocks. The technician gave absolutely no thought to the lay of the land, nutrient concentration at the watering trough, additional paddock breakdowns for more intensive grazing management or the seasonal aspects of the operation.

The total cost of this Neanderthal design? About $10,000. I freaked out. And this friend was all excited because with the government paying 75 percent of the cost, he was only going to have to pay $2,500. That amounted to $800 per acre just to get fencing and water. Compared to our $17 per acre for a highly flexible system, this design and cost was unspeakably horrid.

Allan Nation, editor of *Stockman Grass Farmer,* often points out that really profitable farms have a threadbare look. Their money is going into things other than pretty picket fences, barn cupolas and painted tractors. And to be sure, when you visit us, you will see a farm with a threadbare look. In fact, we have no yard fence; the barn is unpainted and the equipment is practically nonexistent. But it's making good money, and yours can too if you keep things in perspective.

Remember that most of the things we associate with wealth are actually the depletion of wealth. Luxury toys and fancy houses simply drain the bank account; they do not add to it.

With all that said, let's tackle some of the big cost factors in a typical farming enterprise and see if some alternatives exist that can pare these expenses down.

FERTILIZER. What conventional agronomists call "plant food," this is generally one of the largest expenses on any farm. Of course, in the sustainable or ecological agriculture community, we do not use the term "plant food" because we want to feed the soil, not the plants. The whole notion of feeding plants assumes that the soil is a relatively inert material that just holds up the plant and we are putting into it stuff to feed the plant, almost like hydroponics.

One of the reasons I am militant about this farm cost is because seldom is it the weak link. In most cases farm profitability has nothing to do with fertility. It has to do with management and/or marketing. I've listened to more agronomist speeches than I care to remember dealing with the need for soil testing and applying fertilizer. All of us have only so much money and so many hours in a day. We do not have unlimited resources in either category. Our decisions, therefore, are always judgments on priorities.

The question is not whether or not we are going to apply fertilizer, but rather how much do we have time and money for this year, and of all the different types, which ones will be the best buy over the long run? Anyone who can afford to apply all optimal nutrients in one shot probably has way too much money to ever make proper farm business decisions.

At the risk of being misunderstood on this point, let me just say that more often than not, the weak link is not fertilizer per se, but rather some other facet of the farm. For example, while you are out pulling soil samples, putting them in neatly labeled boxes and paying for a laboratory to tell you what to put on for the crop you want to grow, you could be finding out this particular crop is not the one you should grow anyway.

If you do want to pull soil samples, use an independent, nongovernment-connected laboratory that adheres to a philosophy of soil management most similar to your own. A lot of different ones exist. Just so you realize that this is quite an inexact science, let me articulate just some of the options available.

- *Classic chemical.* This is the typical land-grant college soil test. Apply so many units of plant food to grow X crop. These vary dramatically enough based on time of year you pull the sample that if you pull a sample in May and one in November from the same spot you would think they came from two different farms. Newman Turner in his classic book *Fertility Pastures and Cover Crops* explains in great detail the whys and wherefores of this method's pitfalls. In short, he pretty much labels the whole thing a scam.

- *Ionization.* Espoused by the great thinker Carey Reams, this testing focuses on ion exchange capacity and cationic and anionic relationships within the soil. It is used successfully on many sustainable farms.

- *Mineralization.* A system that stands the test of time, most of today's leading ecological farmers believe in the importance of proper minerals. The true blue loyalists in this camp would say that nitrogen, potassium and phosphorus, the classic NPK components, are essentially out the window. A renewed interest in mining ancient sea deposits both in the ocean and in dry areas like Utah and Nevada is making numerous mineral-based products available.

- *Biodynamics.* These folks adhere to special preparations that capture and focus both earth and cosmic forces, unleashing them into plant and soil energies. While this may sound like the most esoteric of all the options, it has plenty of duplicated research to show efficacy and many avid adherents with varying degrees of loyalty. Some people use all the preps while others use only the major ones. Biological transmutation and the "essences" permeate biodynamic conversation.

· *Classic organic.* This includes most large-scale composting operations and organic fertilizer labeled with extremely low NPK numbers like 1-1-2 or 2-1-2. The reason the numbers are low compared to acidulated materials is because this school of thought encourages soil life to release the NPK already there. This is in dramatic contrast to the whole NPK paradigm, but USDA labeling laws require fertilizers to carry an NPK rating on their labels. The low ratings do not suggest impotence. Rather, they suggest a rigid labeling standard that does not recognize the value of other ingredients that act as catalysts or synergizers.

· *Radionics.* I use this term loosely to describe all the low-level energy ideas, from high tech frequency machines to classic towers of power. These folks encourage broadcasting certain frequencies at plants, including certain kinds of music, which will stimulate health, stomata reception and other biological behaviors.

· *Carbon.* Perhaps best popularized by Leonard Ridzon of NutriCarb development, this idea puts carbon rather than minerals on the pedestal. A common phrase has to do with negative and positive spin. These folks trace their roots back to Ed Faulkner, author of the classics *Plowman's Folly* and *A Second Look*, in which he describes his experiments showing that plant material alone is enough to produce and maintain fertility.

· *Foliar feeding.* Although this may not fit perfectly in the soil philosophy category, I am including it because some proponents actually prefer this as the most efficacious way to deal with plant nutrition. Since the leaf stomata can take in nutrients faster, and some would argue more efficiently, than the roots, both macro- and micro-feeding can be done with foliar sprays. The healthier plants then add these nutrients to the soil through decomposition. Commonly, proponents use this technique for supplemental feeding while soil fertility is being built.

While this may not be a comprehensive list, you get the idea that soil science and plant nutrition are quite subjective. To be sure, many of these are not mutually exclusive, but overlap in several areas. I would encourage you to read about these notions and see which ones make more sense to you. My main point is to disparage the idea that the first thing you need to do is get a soil test because it will give you an objective measurement of soil health and a directive for what you ought to do.

Remember too that the faster you try to rectify whatever deficiencies you perceive to be in the soil, the more cash it will take. For example, healing the land with truckloads of organic soil amendments is much faster, but far more expensive, than healing the land with multi-species controlled grazing, composting and animal manures. The problem with the fast approach is that you run ahead of your cash flow. Many profitable businesses go belly up because they run out of cash.

It is better to sell a small crop of vegetables, fruit or cattle at a retail price than to produce twice that amount — way more than you can sell at premium prices — and end up dumping the other half at fire sale prices; or, worse yet, feeding half the veggies to the pigs. In the early years when our fertility was low folks would ask me why we didn't fertilize in order to double our cattle production. My response was that our market was only as big as what we were producing, so why would I want to produce more just to give it away at the stockyard?

Now if you are in the commodity business, selling wholesale, you need not worry too much about the market because no one producer will have any affect on the price of anything. But very few commodity opportunities exist for beginning farmers, in my opinion, and it therefore behooves us to realize that perhaps the worst capital investment we can make is to up our production beyond our market.

Most of us in the profitable end of ecological farming simply apply every bit of organic matter we possibly can. We scrounge whatever organic materials we can find in the area, try to maximize our own manures, and watch the fertility slowly come up. We then plow income into better machines or materials.

The bottom line is to let your market drive the production. Fine tuning your fertilizer program will come as you see a need to grow more product per acre, or per square yard. But be sure you first exhaust other ways to get soil amendments cheaper.

BUILDINGS. What is a farm without a barn? But what kind of barn? Every farm, even if it's a glorified backyard market garden, needs some sort of shed to store tools, equipment and materials. You need a shop area with a vice and workbench. But I am constantly amazed at the Taj Mahal buildings people build.

Even alternative agriculture magazines are full of steel building kits, post and beam plans, and pole structures. You can buy prefab or precut almost anything. People actually put up farm buildings for $10 or more per square foot. I think this is one of the defining factors that separates those who really want to make a living farming and those who don't.

You see, a lot of folks will say they want to make a living farming, but then put up buildings and buy things that actually say: "I want to be somebody, a land baron, perhaps. I want to be a member of the landed gentry, a magnate. In order to prove to my neighbors that I am successful, I need to put on a certain appearance." And the first thing you know, this "pretty little place" is what Allan Nation calls a "land yacht," with a great big money hole in the bottom. The buildings will never pay for themselves. Oh they are pretty, to be sure, but they will never generate enough cash to make economic sense.

I am not opposed to aesthetics. I am opposed to being concerned about form over function and then blaming "farming" for not being profitable. Farming gets blamed all the time for being unprofitable when what is the real culprit is people's desire to keep up appearances and to put on airs. This is insidious because most folks just buy into the accepted way of doing things without ever having had the privilege of being exposed to cheap, functional alternatives.

Perhaps everyone should travel in some third world coun-

Photo 30-1. *A-frame chick brooders cut our brooder housing costs in half because the roof is also the side. Sitting on the ground, these floorless huts can be moved by hand to new locations. Floor space is really all the chicks need.*

tries and observe their building techniques to just gain an appreciation for how much can be done with limited resources. If function is truly your aim, rather than form, you can be under roof for amazingly little money.

We've built several large sheds — I mean thousands of square feet — for 50 cents a square foot. Here's how. We go up in the woods and cut locust poles. They don't have to be perfectly straight. Every area of the country has some sort of rot resistant native timber — tamarack, hedge apple, cedar. We dig a hole and put those in the ground. Then we go up to the woods or a neighbor's and bring down any kind of fairly straight poles for girts. Using a chainsaw, we square the ends and bolt these girts onto the poles.

Using other fairly straight poles of any type, we put up rafters, notching the two ends so they fit snugly onto the girts. These can just be nailed with large spikes. The only milled lumber we really need to use is for bracing from the poles to the girts in typical pole building fashion, and for the purlins across the rafters. In the early years, we tore down three barns in the neighborhood in exchange for all the lumber. We burned what was unusable but had piles of excellent oak dimension lumber left over.

It doesn't matter if these boards are still full of nails. We pound in or pull out what we don't need and nail them up there. All we have in a project like this is our time. Just like you, when we started out, our most crucial need was to keep cash costs down. A friend of mine accumulated a nest egg and then bought a homestead,

Photo 30-2. *A 32-foot x 140-foot awning takes shape, built from pole timber. Some people talk about board feet in their buildings: we talk about cordwood. This is typical in 90 percent of the world and sheds water just as well as a structure costing ten times as much. The animals really don't care if the rafters are of dimension lumber or not.*

which he refers to as his "bank account hemorrhage."

Once you really start making a profit and have money to burn, when your time is far more valuable, then perhaps a "store-boughten" building is okay. But we still build this cheap way. It is perfectly functional, stronger than anything we can buy, has a wonderful wavy, rustic look and still costs only 50 cents a square foot. Many of the old farmhouses in our area have poles for rafters. Flatted with an adz, these pole rafters are just as straight and strong as the day they went up — 200 years ago. Maintaining the full roundness of wood, the full arc, actually gives twice as much strength as the same size squared, dimension lumber.

We mill our own logs on a friend's nearby bandsaw mill. We barter meat and labor for unlimited as-needed use of this wonderful machine. If you cannot finagle a similar arrangement, a bandsaw mill is an excellent investment. For the cost of just one small outbuilding, you can pay for one of the lower-priced models in materials savings alone.

Another cheap building coming on strong is the hoophouse. We have two 120-foot x 20-foot hoophouses that cost only $1 per square foot. We ordered the galvanized metal hoops from a greenhouse manufacturer, pounded the 3 ft. column pieces into the ground with a sledge hammer (the specs call for concrete, but these things have gone through a hurricane unscathed) and had the buildings up in just a couple of days. These are being used all over the northern part of the country and Canada now for winter livestock housing, especially for hogs. Deep bedding and sunshine are an unbeatable combination.

Many different kinds of skins (greenhouse covers) exist. If you build a hoophouse for equipment storage, you may want to consider silvered poly. But for winter equipment storage, a light-emitting skin works great for heating up a diesel engine on a winter day. For a sunlit house, I prefer webbed poly rather than plastic because it is much stronger. Although the webbed material is a little more expensive, I think it's worth it.

Compared to conventional structures, hoophouses are extremely cheap, durable and functional. True, they do not look like the pictures of all the equipment storage buildings in the farm magazines, but they are often 500 percent cheaper and just as functional.

We have built several structures on skids. A huge cost of any structure is the foundation. If you do away with the foundation, the building is not only cheaper, but also portable and therefore more flexible. If you don't want it where it is, you can always move it to a new location. We've built three 12-foot x 20-foot buildings on four locust poles for various purposes. We call them skid houses. They can be placed over an area on which you may want to plant garden. After a year, the ground under these buildings is tilled up beautifully by earthworms and the sod is completely gone. We always put in a couple of skylights to make sure plenty of light gets in. Corrugated steel roofing for both roof and walls is both functional and strong. Quite versatile, these buildings will probably last 40 years before the skids begin to give out.

If you think about it, 40 years is about as long as a structure should be built to last anyway. That is a generation. Chances are the next generation would want something a little different or would want to use a different material. This way you get a building you can pay for and it doesn't imprison the next generation in today's models.

I have a friend who goes to the county landfill whenever he wants to build something. The folks there have gotten used to him coming and let him freely scavenge whatever he wants. He gets all the lumber and plywood he wants. It's kind of a joke now, that if you want to build something, just let him know and he will find the material.

Scrap metal yards are wonderful as well. We have two in our county where, if we take the time to poke around for awhile, we can find all sorts of pipe and I beams, angle irons or square tubing, for girders, columns or whatever. While I prefer wood, sometimes you can find a great deal on scrap metal.

A whole world of cheap buildings opens up as soon as you

get over the notion that a structure must look like the ones in the farm magazines. This battle, like so many others, is really fought between the ears. If function wins out over form, you can enjoy a lifetime of merriment in the "scrounger" society.

So far, I've limited this discussion to farm outbuildings, but I think it would be good to touch on domestic living. Again, I know too many people who bought their acreage and first put $50,000-$100,000 into a new house. That effectively spent their nest egg and then they had no capital left to get production items. No matter what you do, some capital will be required. The more you can shepherd your capital, the longer you can survive those lean years that accompany any fledgling business.

One of the quickest, cheapest ways to get under roof is in a used mobile home. Remember, I'm not talking about forever. I'm talking initially, to get going and put all your capital into the business. You see, your home does not generate income. First you put your fields in order, then you build the home.

I am not going to belabor all the potential alternative home building techniques because plenty has been written about that in other books. Just as Teresa and I lived in the attic for our first seven years, you may need to live in less than accommodating circumstances initially too. If you are not content where you are, you will never be content anywhere. Nothing — money, material, house — will make you happy if you are not happy right now. Home is where the heart is.

Work on being happy pursuing a dream; you will get there a lot faster than you will if you begin putting stipulations in the way. Some of the most common are, "I'll pursue this farming dream only if

- "we can keep both cars."
- "we can have a bedroom for each child"
- "I can have a John Deere tractor."
- "we can go out to eat one night a week."
- "the kids can fully participate in soccer and little league."
- "our place looks respectable."

- "our parents (or whoever) approve."
- "we don't need to adjust our lifestyle."

This is just a smattering of the kinds of issues you will need to address. My goal is certainly not to discourage you, because I deeply believe that for those willing to follow the guidelines in this book, a fantastically rewarding life is waiting for you. But I also get tired of hearing people say: "Well, we tried it and it didn't work." They are quick to blame farming, but when you go for a visit the yard is festooned with several vehicles, a satellite dish and a factory-made storage building complete with John Deere riding lawn mower.

A lot of foolish things have been purchased in the name of farming, just like many a fool has done evil in the name of religion. When I visit the reconstructed homes at Jamestown and Plymouth Rock, the Algonquin homes or the cliff dweller's caves, I have a hard time "appreciating" the energy sucking, anti-solar, indulgent stick houses going up in developments all over the countryside.

Anyway, we can build more for less if we put function over form. Doing so shepherds much needed capital for tooling, equipping and stocking the farm.

MACHINERY. Use ingenuity first; buy machinery second. Go visit some successful small farms to see what minimal equipment they have. Here again, if you need a tractor, that doesn't mean you need the top of the line. If all you need is something to pull things, you do not need two live hydraulic remotes.

Do not buy any more than you absolutely need. Here again, most farmers put no value on their time. When we began large-scale composting, our neighbor came down in his big John Deere and cleaned out our hay shed bedding for us. He put hardly any value on his time because farmers are used to not getting anything for their time anyway. I'll admit that depending on someone else, having to coordinate your schedule with a neighbor's, especially on projects where timeliness is important, can be frustrating, but it is not as frustrating as having to commute to a town job because you overspent on farm items.

Photo 30-3. *A farm can't have too many trailers. They don't require additional engines, they are portable, and multiple-use. Homemade trailers are cheap and functional.*

Photo 30-4. *Our 1966 3/4-ton four-wheel drive Chevrolet with dump bed and high-low four-speed transmission is the kind of machine that costs $1,500 but is worth its weight in gold. It might not be pretty, but it is perfect for beating around the farm. And nobody would think of stealing it.*

We went ten years without owning a garden tiller. We borrowed from neighbors and friends and limped along. Sure, it was inconvenient at times, but that freed up a lot of cash to do other things. Now if we were in the market gardening business, we certainly would have owned one. But we were not. For us, it was much more valuable to put that several hundred dollars into a couple of professional quality, dependable chainsaws.

Concentrate on putting your money into things that will actually pay for themselves on the basis of cash sales, not things that are convenient. We have a fairly large lawn around the old farmhouse, but we still push a lawnmower around, for example. Of course, I'd be happy to do away with the lawn, but I'd better quit on that one before Teresa begins reading over my shoulder.

Now just a word about quality. Almost every manufacturer has a top of the line and a bottom of the line for tools. For example, some chainsaws are built to last 50 hours. Those are the ones sold at the local urban hardware store. They are cheaply priced and cheaply made. Often the most expensive one is overpriced. I encourage getting items priced at about the 75 percent range, especially things like chainsaws, pruning shears, shovels, spades, hammers, lawnmowers and battery chargers. Often the most expensive tool over the long haul is initially the lowest priced. Buying cheap can make you go nuts with broken parts, repairs and poor function. Really good tools are a joy to use.

I remember the first time Teresa got me a Smith & Hawken garden spade. You might have thought I was The Diggingest Dog — you know the children's book by that title? That thing was a pleasure to use. It was balanced perfectly, was heavy enough to penetrate the ground without me having to jump up and down on it. In every way it was superior to the cheaper hardware store models. Don't buy cheap stuff.

Wait until you just have to have something before you buy it, but when you do decide to buy, get a good one. You will save money, time and frustration in the long run by getting a dependable unit. The bottom line: impress people with how much you *don't* have, rather than with how much you *do have.* Doing more with less is a great goal.

FENCING. Although this may not be a big issue in a vegetable operation (unless you have a large deer population), it certainly is important in a livestock enterprise. Here again we can build something that looks like a rhinoceros fence or we can build something more appropriate to the stock we're controlling.

As a general rule, local native rot-resistant wood is cheaper than pressure-treated posts. Most rot resistant varieties will also split nicely, allowing you to make two or four nice posts out of a 6-inch diameter pole. Sledge hammer and wedges work fine.

Fences fall into two categories: physical barriers and psychological barriers.

Physical barriers need to be strong enough to hold against the rubbing and pushing of the stock. A boundary fence should be physically strong. This type of fence will not need daily maintenance and should last 40 years.

Plenty of fencing contractors will vie for your dollars to build a fence, but if you do the work and utilize posts on your own prop-

Photo 30-5. *A simple Speedrite 12-volt fencer and deep cycle battery, one strand of wire and rebar stakes show the elegance of electric fencing.*

erty or a neighbor's, you can get the job done much cheaper. I have worn out many a posthole digger. An energetic man can dig a hole and put in posts at the rate of 4-6 per hour, depending on how rocky, rooty or dry the soil.

For boundary fences I am partial to webbed wire 6-inch-stay field fence rather than high tensile, although some farmers would vehemently disagree. Try a little of both and see which one you like. I see people routinely replacing high tensile with webbed wire. In some areas it seems to be the other way around. Take your pick.

Anyway, the big cost is not in the wire but in the labor and the posts. You can put in 20 posts in a day, by hand, with a digging iron, a shovel and a posthole digger. That's enough for a roll of fence, which goes 330 feet. You do not need this kind of fence internally. Of course, if you're raising something exotic like fallow deer, bison, alpacas, ostrich or llamas, you will need a better fence than this. Make sure you have deep pockets before you start raising these critters.

Fortunately, the animals that most hurt physical barriers also respond the best to psychological barriers: electric fence. Cattle and swine are the easiest to control with electric fencing. Although each responds well to a single strand at nose height, we like to use two strands for pigs simply because their rooting can easily cover up a low strand. This gives a double protection.

I will not go into all the nuances of electric fence construction here because plenty of material is available elsewhere on that subject. My point is that minimal electric fence is cheap to buy, cheap to build and works great for cattle and hogs.

This brings us to sheep and goats, the hardest common domestic animals to keep fenced. They will respond to electric fence, but it must be multi-strand with a killer spark. Any fence that will keep in a sheep or goat will control cattle. Although the production of meat or milk per acre is higher with sheep and goats, the downside is this costly fencing situation. We are experimenting with movable corrals for sheep and goats, and these offer some possibilities for cost cutting. But for a start-up farm, it might be best to stay with cattle to go through the learning curve before launching into sheep and goats. If you will learn on cattle, then your fencing can be extremely cheap and minimal. After you gain confidence you can move onto something else.

Chickens can respond to electrified netting if you clip one wing down to the secondary feathers to keep them from flying over. Doing this does not hurt the bird — it's just like clipping your fingernails. If you draw blood, you're too far down toward the wing. I use a heavy pair of scissors and it goes fast.

Permanent poultry netting around chicken yards should be 6 feet high. We've gotten along with a 50-foot span between posts by stretching a 12-gauge smooth wire on the top and one on the bottom with a winch-type ratchet available at any fencing distributor. Using J-clips like Daniel uses to build rabbit cages, we attach that 2-inch-holed poultry netting to those two stringer wires and it works quite well.

Obviously if you do not have livestock, do not build a fence just for "pretty." That is an incredible waste of money. Except for a corral, I really can't see any use for a wooden fence of any kind, except to spend money.

You can extend the life of poor boundary fences for years using offset insulator and electric wire. The cost of those 12-inch offsets may make you choke, but they can sure get you up and running fast. Replacing fence is a time consuming, money-consuming task. Don't build anything you don't really need.

FUEL. Stay away from on-farm tanks. Yes, they are handy, but you will pay significantly more for fuel than if you just take cans into town and get gas or diesel at a cut-rate filling station.

I know you would think that buying 200 gallons at a time to fill a bulk tank at the farm would be cheaper, but once you pay the hauling bill you will usually pay 10 cents a gallon more than you would at a regular gas station. In addition, you have a pump to maintain.

We keep several plastic 5-gallon gas cans and several diesel cans. When we go to town in the pickup we just take these along and fill them. Not only can we tell real easily exactly how much fuel we have, we also save money.

VETERINARIAN. This may sound harsh to some folks, but I'm going to launch into it anyway. You are running a business, not a hospital or nursing home. If it's ethically acceptable to eat meat — and of course I believe that it is — then it is also acceptable to put a price tag on animals. That means we do not go to heroic efforts to save an animal like we would a human.

Humans are not animals and animals are not humans. Only humans are created in the image of God. Neither did He make of animals a "living soul." While we certainly can get attached to animals, we do not throw cash into a sick one beyond its inherent value.

Certainly there are times when sentimental value comes into play, but that is not farming. That's personal recreation or entertainment. It has nothing to do with farming, even as important, noble or worthwhile as it is.

What is a chicken worth? $2? $3? $5? It sure isn't worth a vet bill — unless of course the whole flock looks sick. If you will follow the husbandry practices outlined in this book as well as my others, you will not have enough sickness to justify hiring a vet.

If the average commercial livestock farmer had our vet bill, each state would have maybe two or three vets. We have a saying that when we have a sick cow we can call the vet and pay $100 and bury her in two days, or we can do nothing and bury her in one day. I'm not suggesting that veterinary services are not important, but I will say that 90 percent or more of the time heroic practices and invasive procedures do not pay off. Furthermore, most would never be necessary if farmers practiced good husbandry.

That includes calving in late spring, after the grass is growing, making pasture the centerpiece of their diet, giving them kelp as a mineral rather than feeding them steroids, and *not* feeding cows chicken poop.

Whenever you are dealing with living things, they die. I heard a great sheep farmer one time say that anyone who has never lost a sheep has never raised sheep. Even the best farmers lose animals occasionally. Those occasional losses will occur with or without a veterinarian.

This discussion, then, has two sides. The first is to realize that animals will die once in awhile despite anything we can do. It comes with the territory. The second is to only use veterinarian services when it will pay. For example, about the only time we ever call a vet is for a calf delivery that is beyond my capabilities. I am not going to lose a calf and a cow to save a $50 vet bill. If the vet says the only way to get the calf is with a C-section, however, I'll thank him and send the cow to the butcher in the morning. Surgery is expensive, and without proper sterile working conditions the chance of success is only about 50 percent. That's too big a risk for me. Better to burger the cow.

That may sound harsh, but it is not a moral issue in my opinion. Once you get over the "poor animal"/Bambi syndrome, it simply is a practical matter. Do I pour more money into this animal or do I go ahead and salvage what I can? Anyone who knows me knows that I love animals. I can get real emotional over a sick animal, especially a pet or an "institution," like a special bull, rooster or cow. But there is a cycle to life, and it includes death, just like it will for me. I don't want a bunch of heroic efforts expended to extend my life a few months either, for that matter.

We need, again, a balanced view of how our animals fit into the great scheme of things. By all means we need to be well-read and well-informed not only about good husbandry practices but also about good remedial techniques. Remedies that work on humans also work on animals: chiropractic, herbs, homeopathy, acupuncture, and supplements. Plenty of good books explain both current and old-time remedies or supplements, like feeding charcoal to pigs or giving laying chickens a wood ash dust bath to eliminate lice. Please don't think for a minute that I would minimize these techniques to either maintain health or cure sickness.

But after we've done what we can, there comes a point of diminishing return. In my experience, if an animal really wants to live, it will have unbelievable resiliency. We once had a cow get stuck under a fence while having a calf. She was paralyzed. We brought her to the house and carried her hay and water for thirty days. The next morning she stood up, staggered around, and within a couple of days rejoined the herd. Of course, we culled her that fall, but she sure surprised us.

Of course, we've had others go the other direction. Back when we foolishly calved in January and February along with everyone else, we had a calf born in an ice storm one night. I found it about three hours after it was born. I brought it into the barn, mother following, and felt for a pulse. The calf was cold as ice; its mouth was cold and it looked dead, but the heart was still beating. I milked the mother and fed the calf the warm colostrum through a stomach tube. The next morning the calf was up and nursing on its own. About three days later it started limping. I caught it and to my horror could not feel any circulation in the bottoms of its legs. Three legs had one little strip that felt warm. We bandaged the legs and penned it up where we could keep the little fellow clean. I milked the cow a couple of times a day and bottle-fed the calf. We waited to see if at least three legs would have enough circulation to heal. After a month, it was apparent the little fellow would lose all four legs, and we mercifully ended its suffering.

I could tell story after story after story about doctoring animals — some successful and some not. Every farmer who loves animals — and most do — can tell similar stories. But most of the time these emotionally draining and time-consuming problems are caused by a lack of proper husbandry models, which include culling old animals before they get sick or unproductive.

We had an old Angus cow, #11, who produced a calf every single year for 16 years. She also had a big, strapping, healthy calf. I never had to assist her and she had tremendous maternal instincts. I could never eartag her calf because if I came within 10 feet of the little booger she would kill me — and she did try several times. I've gotten aerial lifts several times thanks to the head of a protective cow. Finally one year her calf was one of the smallest ones and we knew if we waited one more year, we would probably have calving trouble. As difficult as it may be emotionally, we must get rid of those animals before they become a greater liability. If this seems unloving or inappropriate to you for any reason, give yourself a few years on the farm and you will see the wisdom in it.

A fellow in our county runs a huge sheep operation. He's a good farmer, as conventional farmers go. He told me the most profitable way to run a sheep flock was to not check on them at all during lambing. Yes, lamb in the late spring, not in the snow and cold, but don't check on them at all. In the fall, go bring them all in. Some will live and some will die. You'll go bankrupt trying to save every one and it tears on your emotions too much.

A bleeding heart farmer will soon be a bankrupt farmer. Animals are for function, not just form. Be ruthless in culling poor stock and you won't need to deal with as many problems. I've talked to countless folks who lament putting $500 in a ewe. Forget it. Sometimes you just have to let the buzzards feast. That's better than letting them feast after you've spent several hundred dollars.

One place where veterinarians are extremely useful is in diagnostic work. If you want to know what your parasite load is, for example, take some fecal samples in to the vet. Find out what's going on and solve it; don't just use the pharmaceutical recommendation and leave the animals in the mud.

Lest anyone think I am some kind of ruthless animal hater, I implore you to come and visit us. You will see the shiniest, slickest, healthiest livestock, the most contented animals you have ever seen. And you'll probably see the lowest vet bill per gross income dollar of any farm in the state.

As we draw this chapter to a close, I trust that you have a new appreciation for just how easy it is to spend money in the wrong places, and that you have developed a greater "can do" attitude. All the big ticket items that your farming acquaintances will tell you about to dissuade you from farming need to be seen in the light of what is in the farm magazine advertisements. That cannot be your mental picture if you are going to make money farming.

Instead, you will see threadbare functionality and a fat bank account. When that happens it will make all the disparaging comments meaningless. Your joy will be in what works, in having beat the odds, in having found and lived your dream.

Labor

"I can't get no help." I wish I had a nickel for every time I hear farmers say that. This lament is pregnant with meaning. For sure, our modern culture denigrates blue-collar work and glorifies white-collar. Certainly our welfare state eliminates the many rural farmhand types that farmers relied on a few decades ago.

But these dynamics notwithstanding, this typical farmer lament is a telling commentary on what has happened in agriculture. Devoted to getting warm bodies off the land, the entire agricultural sector has quite successfully relocated country people. Farmers bought into the industrial mentality, trying their best to eliminate the labor requirement, and are now reaping the awful harvest.

Because he has become a feudal serf to the vertical integrator, the fertilizer dealer, the equipment manufacturer and the Madison Avenue advertising agency, the farmer no longer even earns minimum wage. I can't believe how many farmers I've heard actually admit that they are not really trying to earn money. It's almost as if earning a decent living is an evil intent. It's okay for a doctor or lawyer, a college professor, plumber or electrician, but just don't expect a farmer to have the same ambition.

I believe a farmer's ability to attract labor, or "help," as we say it, is directly linked to his own feelings of self-worth. When a farmer thinks he is not valuable, he will not place value on his help.

I enjoy getting other farmers to help me do things because I never cease to be amazed at how cheaply they work. I bought some pigs the other day from a guy who told me he was not trying to make any money at raising hogs. I gave him way more than he asked because I want him to stay in business. I value his work because I value my work. I'll bet nobody has ever paid him more than he asked.

As a farmer, you need to understand that you are a professional businessperson. I'm known in conference speaking circles as the "coat and tie guy." That is not because I'm ostentatious or blessed with a high-falootin' wardrobe. You become who you perceive yourself to be. If you think you're just a lowly hayseed out here bending your head to the hoe and that's your status in life, then that's exactly who and what you'll be.

But if you are a professional, think like one, dress like one and act like one. And acting like one means having enough self-respect to refuse to work for minimum wage and to refuse to let anyone who works for you do it for minimum wage.

In the early years while Teresa and I were still struggling desperately to make ends meet and before Daniel and Rachel were big enough to help out too much we needed to hire help occasionally. I always paid at least $10 per hour. Guess what. I never have trouble getting help — good help. I can even elect to be choosy.

You can hand-pick your help if you're willing to pay enough. I would rather pay a fellow $10 per hour who puts out than $4.00 per hour to a slacker. Measured by any standard, whether maximum output for minimum input, emotional satisfaction or total work accomplished, the higher pay for higher caliber will always be your best investment.

A slacker is not a good investment at any price. We've had some of those too, and I don't care if they work for free; they'll never pay their way. Life is too short to put yourself through the frustration of a dud. Concentrate on getting good help and be willing to pay for it.

When I've wanted help, I go for specific people — the best around. I don't go for the guy who has nothing better to do. I go for the fellow who is already busy. You know the old adage in volunteer organizations about "if you want something done, ask the person who is busiest." They seem to have a knack for getting much accomplished. If the best a guy can do is sit in front of the TV sipping soda, he's probably not the best worker.

I had a fellow call me once who was interested in doing an apprenticeship. My first question was: "What are you doing now?" He responded: "Well, I graduated from college three months ago and have just been sitting on my rear in front of the TV ever since." Can you imagine? Needless to say, we did not have an opening.

So where do you find energetic, conscientious help? My first suggestion is to link up with your local home-school support net-

Photo 31-1. *Rachel plants cabbage sets as well as any adult. She has a green thumb, has learned how to do it from Teresa and me, and is an energetic, creative gardener.*

work. We have found local home schoolers to be a cut above average in every respect. Since they are not dependent on peer approval for everything in life like normally schooled kids, they are happy to work, even at unpleasant jobs, and not worry about what friends may say.

They tend to come from good homes where character virtues are taught. After all, Mom and Dad have modeled personal sacrifice in order to home-school their children, and the children in turn model those same qualities. Often home-school families have several children and are strapped for cash. Some extra spending money comes in real handy. Also, since the young people are not in school every day, they have physical, mental and emotional energy to devote to your task.

Obviously, any industrious neighborhood young people, if treated fairly and paid exceptionally, will perform acceptably. Geographic nearness is a real asset. Pay better than you promise and you'll be surprised what will open up.

It may be helpful to reprint here a piece that my dad kept under his desktop for years. It comes from a magazine called *Practical Supervision.* It offers several questions to help evaluate employees:

1. What does he admire? Determine a man's heroes and you have a good clue to his motivation.

2. How does he spend his free time? Where he goes, what activities he pursues, whom he associates with are good indications of his educational and cultural interests.

3. How does he spend his money? What he buys and where he lives tells a lot about his sense of values and suggests a lot about how he would handle your assets.

4. What makes him laugh? And whom does he laugh at? Answers to these questions reveal significant insight into a person's emotional stability.

5. What does he read? Knowing about a man's literature can tell you about his aspirations as well as his escapes.

6. How does he handle setbacks? Reaction to adversity discloses an individual's sense of self-worth.

When we began our weekly delivery service to restaurants we did not want to do it ourselves because we enjoy the farm too much. As we began looking around for a subcontractor we did not look for a cheap driver. We wanted someone who had a stake in the success and growth of the enterprise. We pushed ourselves for a year and built the route up to about $100 per half-day delivery run.

At that point, we felt like it was economical enough to attract a high caliber person and actually had to narrow the field from half a dozen folks who had heard about the opportunity. The fellow who is doing the delivery now often earns more than $200 for the half-day drive once a week. I've actually had people tell me I'm paying too much: "He's earning more than you!"

Which brings me to the next part of this discussion: commissions. As a general rule, wages and salaries are unrewarding. They do not stimulate excellence. I think that is true in all sectors of society, but I will confine my remarks here to the farm. Nothing encourages creativity and a zest for the work like rising and falling with performance. That is the fundamental flaw of hourly or salary pay — it does not inherently reward performance. Performance is fairly subjective depending on what the boss thinks. Is the worker putting out to his full potential, or is he slacking? Is the worker slow because he doesn't understand, or because he's lazy? Is the worker cutting corners, or just being exceptionally efficient? These judgment calls can get sticky for everyone. I am a believer in commissions or "by the piece" remuneration.

When we began our delivery, therefore, we did not want an employee relationship; we wanted a subcontractor relationship with hefty incentives built in to stimulate success. It worked so well that in only one year the weekly delivery share doubled. Do I care if he makes $50 per hour? Not at all.

What we do is sell everything based on an FOB Polyface price. Then we add on a delivery surcharge based on pounds that is graduated to reward volume purchasing. It looks like this:

GROSS WEIGHT (IN POUNDS)	**DELIVERY CHARGES** (CENTS PER POUND)
0-50	33
51-100	30
101-150	27
151-200	24
201-250	20
251 and up	17

What this does is

- encourages the customer to buy more in order to reduce the delivery surcharge per pound.
- encourages the delivery subcontractor to sell more because the cost is in the trip, not carrying stuff inside to the customer.
- encourages our profit because the volume discounts are not absorbed by us but by the transportation, where it rightfully should be.

We break out the delivery surcharge on the invoice so that the customer realizes that we, the farmers, are not getting the entire amount. This helps to educate the customer as to the cost of transportation. The average consumer has no idea that the farmer gets less from a box of cereal than the cardboard manufacturer. It is time to develop marketing models that educate rather than ignore (the linguistic base of "ignorance").

We don't care if the subcontractor earns even $100 per hour because we're still getting our full retail Polyface price. Everyone wins, and that's the goal. We do not want to create marketing scenarios in which winners necessitate losers. We want all parties to win.

This same idea can be duplicated, with appropriate permutations, in all sorts of ways. For example, don't hire a couple of guys to help you put in hay by the hour. Rather, pay them per bale. Make it so they can earn $20 per hour if they really put out. If they actually do work that hard, you'll be money and work ahead by the end.

If you hire someone to man a booth at the farmers' market, put a commission on the project. The worker gets to keep 10 percent of gross sales, for example. The money you lose will more than be made up by the new demeanor and creativity of the worker, who now has a vested interest in seeing sales escalate.

If you have garden weeding to do, pay by the row-foot, not by the hour. Hourly pay has created millions of lazy people. Except for personal integrity and the displeasure of the boss, the worker has no incentive to perform exceptionally. While I would not say that the only thing that matters is performance, I certainly lean that way. The more we can link reward to performance, the more insurance we have of a mutually beneficial arrangement.

Another type of arrangement is partnering. If you do not like to market, then partner with someone who does. Find an Amway distributor and arrange a profit-sharing option. You can get better than wholesale price but not full retail. The market partner gets a commission on sales. Each person exercises his talent and the result is better than either trying to operate alone.

The same is true on your farm. An elderly farmer can bring on a young person to operate an additional enterprise and in exchange, help out the older farmer with his labor requirements. For example, a beef or dairy farmer could have a nearby partner come in with a pastured poultry enterprise or a vegetable/small fruit business without reducing the parent enterprise at all. In exchange for using the older farmer's machinery and land, the young fellow gives labor. Both parties win.

If you're the young fellow, make the offer to an elderly farmer, or for that matter anyone you know who owns land. I know a fellow who spent a couple of months marketing this idea to farmers in his area until he finally found one who would let him run pastured poultry on his acreage. Now the man wants him to cover his farm with those pastured chickens because the grass grows better due to the fertilizer the chickens drop.

An orchardist should welcome a pastured poultry enterprise among the trees to eat bugs and drops. I am convinced that we have not even begun to explore these potentials in agriculture. For example, enough undigested grains come through the manure on the average 100 cow dairy to support a $10,000 per year egg laying enterprise. A feedlot would be even more profound. You could build a layer house into the fence line of a cattle feedlot and produce eggs for free; the owner would see his fly population diminished as a spin-off asset. These symbiotic relationships are real and only creativity stands between you and making one of them work.

We have eased our labor load with apprentices. In order to attract good apprentices you must have an ongoing, viable farming enterprise. This is not something for start-up folks, but can certainly ease the workload as the farm expands. We provide room and board plus a stipend based on farm cash flow and individual initiative. We try to require apprentices to make a one-year commitment, although we sign no agreements.

The risk to the apprentice that we could kick him out anytime is offset by the risk to us that he could leave anytime. This arrangement builds mutual respect and offers boundaries for behavior and interchange.

Over the years, we have received much free labor from folks

Photo 31-2. *Polyface apprentices Micha Hamersky from Austria and Jason Owen from Virginia enter their cottage with attached solarium. Jason's pickup truck is his first "farming" acquisition. After a year of working alongside the Salatins, these young men have skills to duplicate and refine the concepts outlined in this book.*

who are just curious about what we're doing. One of the best ways to attract help is to do something new, or something different. If your farm is the same as everyone else's, it will not be enticing to someone wanting adventure. At the risk of alienating someone, let me say that this kind of help is a two-edged sword. Some folks can be helpful and others would be most helpful by staying away.

You as the farmer are pushed to keep more people busy, to keep the work progressing and to deal with more personalities. The chances are myriad for inexperienced folks to break a hoe handle, to crack a shovel, to misplace a wrench or break a trailer hitch. The way some handle machinery or tools you'd think they were suicidal.

All of this comes with a price. We had a fellow one time who broke three drill bits in 30 minutes. Thanks for the help.

But as long as you appreciate the expense of the help, the downside, you will come to appreciate overall positive impact these folks can have on the farm. Not just in the work itself, but in ideas and problem-solving. A few different perspectives can help.

An attractive work environment means everything. How many people have a deep-seated desire to go spend a day in a factory chicken house? Not too many. The agricultural models we use can attract or alienate. This goes hand in hand with our own demeanor as well. If we're always down-in-the-mouth about farming, not too many people are going to want to come and spend the day with us. But if we're upbeat and excited, folks will want to come and see what energizes us. A lot more people are attracted to a fire than to a garbage dump.

A significant source of labor for us over the years is our fellowship group. We currently home-church and encourage community interaction by holding workdays at each other's homes. It may mean helping someone move, building a structure or processing chickens. Often on big projects one plus one equals three. When we erected our sales building the whole group came and put it up in two days. Knowing that we have that support network makes us much more willing to tackle otherwise impossible tasks.

Not having a church building with related pastor salaries, insurance, heating and cooling bills and maintenance requirements frees us up as a group to help each other more. Whether the help is in the form of labor or money or both, it strengthens the loyalty in the group to have interconnections outside the meeting time. These business, barter and relationship interconnections encourage charity when potentially divisive doctrinal discussions develop. I speak more charitably to a person with whom I disagree if we have a mutually beneficial extra-church arrangement going on like trading labor or sharing a business opportunity.

When we wanted to automate our chicken dunker an engineer in the group came over and designed the hardware, wired it up, and kept it maintained. The only cost to us was a bunch of thank-yous, a couple of meals and all the free chicken they want to eat. People want to be needed. It is more blessed to give than to receive. Do not overdo the asking part, but if you are trying, and truly carrying your end of the load, you would be surprised how willing people are to help you.

Trading work with friends and neighbors has always been standard farm policy. Pursue this wherever you can. The town job farmer, though, has made this much more difficult these days. However, often the farmer working off the farm has more equipment than necessary. I have a neighbor who works in town and I have often raked his hay for him so it's ready to bale when he gets home from work. In return, I can use the hayrake and keep from buying one. The list of barter arrangements is as long as you want to make it and can involve as many deals as your creativity can concoct.

If you devote yourself to people, holding them in high esteem, farm labor can be a fairly easy problem to solve. But if you devote yourself to dogged independence, farm labor will be hard to come by. The blessings of a good attitude, a positive outlook and putting others on a pedestal will pay big dividends when you need an extra pair of hands to do a project. Commit yourself to people, and they will commit themselves to you.

Accounting

Dad often told the story about his Venezuelan accounting friend who, in a heavy Spanish accent, would tell him: "Señor Salatin, you must poot eet doowwn." As I see it, this is the overriding principle of keeping farm accounts. If you do not write them down, you can never reconstruct them and you never know what your financial picture really is.

No one dislikes accounting more than I do. I would just as soon forget money altogether. I would rather spend all day chopping multi-flora rose bushes out of a fencerow than have to balance the checkbook. Fortunately, Teresa is meticulous to a fault and can make that thing come out to the penny like a professional.

Gifted at setting up easy-to-keep, meaningful record-keeping systems, Dad was a stickler about getting every item of information written down in a way that would make it useful for decision making.

Too many farmers just throw all the receipts in a box and at the end of the year try to reconstruct what happened. Memory doesn't work. Even a small farm has many different aspects. If we are to know which ones are draining money and which ones are making money, we can't just lump everything into income and expense. Each item must be categorized so we can track the progress of that business component.

What follows is a partial list of the categories we use:

Expense

- Broilers
    - chicks
    - feed
    - supply
- Cattle
    - calf
    - feed
    - other
    - slaughter
    - supply
    - vet
- Contract Labor
- Depreciation
- Dues and Subscriptions
- Eggs
    - chicks
    - feed
    - supply
- Electric
- Farm Machinery
    - fuel
    - labor
    - lube
    - parts
- Fence
- Fertilizer
- Forestry
    - supply
    - walnuts
- Insurance
- Interest

Income

- Broilers
    - chicken
    - chicks
    - feed
    - supply
- Cattle
    - beef
    - feed
- Eggs
    - eggs
    - pullets
    - stewing hens
- Farm Machinery
- Forestry
    - firewood
    - lumber
    - walnuts

Expense (cont'd)	Income
Loan	Loan
Pork	Pork
feed	
piggies	
Postage	
Promotion	
Rent	
Repairs	
Salary	
Shop	
Sundry	
Supplies	
Suspense	Suspense
Taxes	
Telephone	
Travel	
Vegetables	Vegetables
Vehicle	

As you can see, we are intent on gathering information. That is the purpose of accounting. All the effort is wasted unless you use it. Every farm will have a different set of categories, but I think this gives the idea of the kind of information you need and how to arrange it in a logical order.

How it actually works from day-to-day is that whenever we buy something we get an invoice. We write on that invoice the proper category so that when we put it in the ledger (we use Quicken software) we don't have to remember what each item was. If an invoice has multiple items that belong to different categories, we go ahead and write those notations on there so we'll remember.

If we purchase an item, like kelp, which will be used for multiple categories, we try to divide it among several categories. It's not fair to charge all the kelp to cattle when we use half of it for chickens. Obviously, this cannot be done perfectly, but at least we attempt to get a fair distribution to the different enterprises. Of course, some things cannot be prorated that way. A mechanic bill for the tractor, for example, would not be charged to livestock. Those kinds of things have their own expense slot with no offsetting income slot.

The whole idea here is, as much as possible, to fairly assess the performance of the different farm enterprises. One of the most common errors in farm accounting is to have high performance enterprises subsidizing low performance ones. You either want to identify the problem in the low performing one and rectify it or just do away with that enterprise altogether.

But if you can't tell which enterprises are profitable, you can't know where to focus your efforts. We don't want any enterprise to carry another; they must all stand by themselves. While this record-keeping may seem tedious, it is invaluable to keeping you on track.

One of the biggest problems in small business is what Dad called "slippage." That is the potential profit that slips between the cracks. For example, Dad had several mechanics as clients, and although they were exceptionally skilled, they seldom really made money. They worked long, hard hours and never seemed to have any real profit. He would go in and get them to keep a time card, accounting for every minute of the day. They started billing their full time on jobs and suddenly saw their income double. They were doing favors, yakking and generally running the business like a hobby instead of a business. I have no fault with maintaining a flexible, low-stress working environment, but it needs to be tightly run enough to make a profit.

We use these figures to determine marginal reaction and gross margins. What do you make on one beef or one pint of raspberries? I like to boil things down to our margin on a single dozen of eggs or a single chicken. By keeping these figures in your head, you can stay encouraged when you're tired. "I just made another $50," you can tell yourself as you prune the apple tree. If all the money is out in some nebulous, jumbled sphere, your efforts will lack immediacy and direction as well.

I know this type of record-keeping is not being done because I see farmers routinely spending $10,000 for a piece of equipment that will not pay for itself in 50 years. If we're going to spend some money, we need to spend it where it will give us the greatest leverage in meeting our goals.

This brings us to the discussion about what to do, what not to do and where to stop. As entrepreneurs, we are tempted to chase every potential customer, every sale. Whatever someone wants us to do, we want to do it. Often this enslaves us to low return items while high return items go begging.

For example, we used to clean all of the gizzards from our pastured chickens. After all, we grew up with the "waste not want not" mentality and gizzards were certainly edible parts of the chicken. The problem was that even with a machine it took one person an hour to clean $10 worth of gizzards. Not only that, but we found many of our customers did not like them. We realized that we could clean $100 worth of chickens in the time it took to clean $10 worth of gizzards. As difficult as it was, as much as it grated on our psyche, we now just throw the gizzards away. Cleaning them is an expense.

People routinely want us to cut up chickens for them. Because we are not automated, it takes nearly 5 minutes to do that, which translates to 12 per hour. To make our $25 per hour minimum, we would have to charge more than $2 extra per bird and people will not pay that. Therefore we just tell folks we do not offer that service. We'll gladly do a couple for them to show them how to do it, but we will not do it for them.

We run into this with our firewood business. Around here, the normal market is $45 per pickup load cut, split and delivered. It takes one hour to cut and load a pickup, one hour to deliver it and one hour to split it. That's three hours in that load and roughly $10 in expense: chainsaw, fuel, and machinery wear and tear. Therefore the return is a little more than $10 per hour, return to labor, excluding the cost of owning the pickup truck. Instead, what we do is cut it and put it on a big pile here at the house and let folks come and get it

U-haul. We sell it for $30 per load, unsplit and undelivered. After taking off $5 for expenses, we're left with $25 per hour return to labor with the same equipment.

What's interesting is that the customer doesn't attach a cost to running out here with the pickup. It's already fueled and sitting in the driveway, so the overhead has already been created. Although we do not get the sales of people who want it split and delivered, we have found a huge niche among people who don't mind putting a little sweat equity into it. And if their time was going to be spent sitting in front of the TV anyway, the labor is not a cost either. Actually, it is an asset because the exercise can replace the local spa dues. Many of them view the $15 per load as significant cash savings.

I don't know about your area, but in our area men are always trying to justify owning a pickup truck to their wives. By allowing them to come out to the farm to pick up their firewood, we give them a reason to own that truck. After all, needing it to keep beer cans in just won't wash with most wives. But if the truck is necessary in order to keep the house warm, they can get by with a $30,000 Dodge Ram, don't you think? We're in the business of meeting the whole person's needs: physical, emotional and mental.

A losing proposition in this area is backgrounding feeder calves. Many cattle farmers hold over or even buy calves in the fall and winter them on grain and silage in order to sell them heavier on the higher spring market. If you will account for the costs per pound of wintering compared to selling and buying back in the spring, you will find it never pays to buy machinery and make silage to overwinter those calves. The cost of weight gain is just too high to compensate for the added overhead.

Knowing when to stop in your value adding is as important as knowing when to start. Not every amenity people are willing to pay for will be worth your effort. Only by keeping a sharp pencil and putting it all down will you be able to determine what should or should not be done.

At the end of each year you need to sit down with these figures and run enterprise profiles to determine which ones are making the most money and which ones are not. Steel yourself against the urge to say: "But I enjoy doing that." Plenty of people have gone bankrupt doing things they enjoy. Let the figures speak, and be courageous enough to learn from them, to make whatever changes are necessary.

People will try to talk you into doing unbelievable things for them. One of their favorites is to get you to deliver. But as soon as you start looking at the cost of transportation and the time away from home, most of the time you will see that delivery is a dead end. Still you may have a friend or neighbor who enjoys beating the pavement and would gladly work as a subcontractor to make deliveries for you. Just remember to allocate all your costs to the appropriate enterprise.

In general, you cannot get too much information and you cannot overuse it. Stay on top of your money and you will find it enjoyable to monitor how the business is flowing. Never trust your memory alone. As Dad's Venezuelan friend said: "Put it down."

In addition to straight monetary accounting, you will want to maintain other accounting type information. Certainly a market gardener would want to write down purchases of seed varieties and volumes, where each was planted and how much square footage allocated to each vegetable. As the season progresses, he would want to record which vegetables sold best in order to decide whether or not to expand or contract the square footage allocated to that variety next year.

We keep close breeding and calving records on our cows. Certainly Daniel's records on his rabbits are extensive. I keep detailed records on the grazing rotation, including how many cow equivalents were in what field at what time. That way we know whether the productivity, measured in cow-days per acre, is gradually coming up or going down. The grazing records also tell us where we started each spring so we can rotate that as well to stimulate forage diversity. We record compost applications as well.

If you're running a bakery, don't just record gross sales.

Record each transaction by type so you know which items sell best. Record, as well, time spent on each item. You may well find out that the item selling best is only returning $5 per hour to labor and one of the poorer sellers makes you $15 per hour. Perhaps you would be better off to accelerate the marketing effort on the higher returning item rather than producing more of the best seller.

You can hardly gather too much information when you're running a business. Decisions are based on the information you have. If you don't know the information, you simply cannot make a wise decision.

In addition to production accounting, you will want to maintain what I call marketing type accounting. This involves a "potentials" list, or some may refer to it as a "tickler file." It is full of information, including mutual friendships or contacts, which will be useful in making your next marketing call to that potential client. Perhaps you do not have enough volume to access this client yet, but you need to be accumulating information so that when you do finally make the contact, you will be able to make a better sales presentation.

Noting on a card that a potential patron is a chef where one of your customers frequently eats can be invaluable to making a favorable first impression. Knowing that a client specializes in exotic fowl dishes can move that prospect up or down on the "to contact" depending on what birds you decide to grow.

You will certainly want to maintain a sales history so that you can look back and see exactly who bought what last year and the year before. Sometimes we have customers send their order blanks back with the following statement: "Just put us down for what we had last year. It was perfect." Obviously if I have no idea what they got last year, I can't fill that order.

Whether you do this on computer spreadsheets, make your own paper charts, or use multi-columned ledger paper is really immaterial. This information is the heart of your business. Accessing who bought what and when is how you chart trends for your business. It helps you form "the ideal customer" profile and other marketing tools you will use as the business grows.

Much of accounting is simply getting organized. That includes keeping a date book for appointments or other important things. For example, when we buy replacement pullets I go 20 weeks ahead in the date book and mark down that those chicks are supposed to start laying then. Invariably, if I don't write it down, 16 weeks will go by and I can't remember when that batch of birds was supposed to start laying. "Sometime soon, I guess" is just no way to run a farm.

Getting organized includes being able to deal with the constant flow of paper that crosses your desk. For some folks, it may mean getting a desk! Jeff Ishee shared the following principle with me: "Every piece of paper should only cross your desk once. Take the necessary action, and then file it." I seem to remember him looking down at my desk and shaking his head from side to side as he said it.

I have instituted one policy that seems to really help keep the clutter down. Anything that comes without first class postage automatically goes in the trash. I don't even open it. If someone wants my attention, they'd better send it first class. I've read recently in marketing books that a mass mailing is more cost effective if you use first class postage and mail half as many than if you use bulk for twice as many. That's something to think about.

Accounting is one of the most important aspects of your farm enterprise because it is the information upon which you will base most of your decisions. Think of it as keeping your personal history. These records are your biography, or at least that of your farm business. The more accurate and meaningful they are, the better your decisions will be. And all of us aspire to make better decisions.

Filing System

We all need a way to store information and retrieve it efficiently. Even though computers are filling much of this function, we still need to keep paper copies of important documents and magazine articles, as well as notes and records.

In order to function, a filing system needs the following:

- No individual entry may contain more than half a dozen items. If any file contains more than that, it is dysfunctional inasmuch as you cannot isolate individual items quickly. If you have to root through 20 or 30 items to find the one you want, the file is too bulky to be used efficiently.
- Files must be flexible enough to be adjusted to keep the number of items low. When the items in a file reach 10 or 12 and you want to subdivide, the system must be able to accommodate additional entries easily.
- All options should be easy to see. No two people think exactly alike. You and your spouse, or business partners, or children will come up with different wordings for headings. In order to be useful, all the options must be quickly scannable in order to facilitate others' use of the system. If the heading is not what you would immediately think, rooting through the entire file cabinet looking at manila folder titles to find the right one is maddening.

- Locating headings should be easy. Flipping through the file cabinet reading headings is not very efficient.

With these parameters in mind, I use a filing system modeled after the ones we always used in debate competition. I will be forever grateful for that experience. We would go into a debate round carrying 3,000 or more individual pieces of evidence on note cards. These were totally unrelated because debate offers a wide latitude in how to accomplish the topic resolution.

In a debate, we had a total of ten minutes, cumulative, to prepare our speeches. An important part of that was finding the information we needed to support our statements. That meant we had to find perhaps a dozen individual pieces of evidence from our pile of 3,000 in about one minute flat. That's a lot of stuff to look through for one little item to support a point.

Between debate tournaments, we would research areas in which we were weak to strengthen our arguments the next time we encountered opponents using similar ideas. This meant that each week we would add to and revamp this pile of evidence cards, which meant changing headings. This whole procedure would be a mess without an efficient, versatile filing system.

What we used was an outline scheme. Rather than writing headings on file box tabs, we created a master outline. Lest you think I'm clever, this system is used universally in interscholastic and intercollegiate debate competition. It is such an efficient system that I use it for all of our filing. Others can follow it and use the file as well. Folks who have a notorious problem "keeping up with things" find this system revolutionary and I think you may as well.

Anyway, the outline articulates, in logical breakdowns and sequence, all the file headings. The tabs, whether they are index cards or manila folders, are simply numbered beginning at 1 and going through whatever.

The outline is simply on regular paper(s) and each outline entry that is a file heading has the file number out to the side. This master outline is the brain, the organizer, of the files. The beauty of this system is that the outline can be changed easily, even completely revamped, without messing with file tabs. At the same time, such changes do not require rearranging the files in the cabinet.

The systematizing of all the headings, in a logical sequence, allows you or anyone else to review all the options in a jiffy. That way if your brain and my brain don't work exactly the same way, you can quickly scan the outline and see how I may have written down the heading you are looking for. In addition, we can make huge changes in the system, even to the point of adding 100 headings, by simply changing the outline and adding a bunch of file numbers. Two outline entries, quite similar but broken down to keep the individual file down to perhaps 6 items, may be right next to each other on the outline but number 3 and number 303 in the file cabinet.

Just so you can visualize what I'm talking about, let me put some of my outline headings down. Here it goes. Headings are on the left; file numbers are on the right.

I. Agriculture
  A. Polyface Inc.
    1. Direct Sales
      a. Beef 1
      b. Poultry 3
      c. Pork 9
      d. Promotion
        i. information sheets 92
        ii. past newsletters 93
        iii. old order blanks 217
        iv. customer testimonials 387
    2. Production
      a. forage
        i. pasture rotation 7
        ii. soil amendments 5
      b. beef
        i. breeding 8
        ii. calving 4
      c. poultry
        i. ration 30
        ii. hatcheries 200

C. Personal/family
   1. home education 52
   2. ministry
      a. fellowship business 134
      b. missionary prayer letters 215
   3. transcript and resumé 53
   4. Salatin family — history 155
   5. House papers — history 162
   6. Christmas letters 215

Are you worn out yet? This is an extremely, extremely abridged outline — the full deal would take 20 pages. But I tried to put in enough that you could get the drift and begin to appreciate just how slick this system is.

Our file goes to about 400 and can be modified easily simply by changing the outline and adding numbers. When a file gets fat, we try to see how we can subdivide it in order to facilitate finding specific things. The old file will stay where it is, but the new items will be moved to the next unused number: the heading will be right in the proper outline sequence.

Nothing is more frustrating or laborious than picking through a file cabinet trying to find something in that hodge-podge of tab notations half worn off or bent over. Obviously our file is much bigger than it was a decade ago. I am interested in more things than I was a decade ago, too. This format allows me to add new areas of interest without upsetting the apple cart.

One big problem we need to address is what to do with magazines. Almost every magazine you subscribe to will have at least one good article, but if you leave it in the magazine and add it to the stack in the cardboard box, it is completely useless. Information you cannot access quickly is essentially useless.

I cut out the articles I want to keep and file them. While to some folks it may sound sacrilegious to throw away all those pretty slick covers, I prefer that to the alternative: a mountain of unusable information. Over the years you will be surprised how much information you can accumulate on a given topic. You can also give them away.

My file is certainly as useful as books. If I want to know something about boron or switchgrass, I can find a pile of targeted information in the file and do not have to try to look it up in books. Although I would never disparage the usefulness of books, often it may be hard to put your fingers right on the page or maybe even the book you remembered reading something in. This file is like a personally accumulated encyclopedia and now, decades after beginning it, is a veritable fount of information.

As you read through these outline headings you may have wondered why I would use certain word combinations, but these are meaningful to me. And in a moment you can scan the list and see all the options if you don't think just like I do.

The outline keeps things in logical, sequential order to facilitate finding things even in narrow subdivisions. While at first glance it may look like a lot of work, it will flow easily as you begin categorizing what you need. Time spent getting organized will more than pay for itself when you need information.

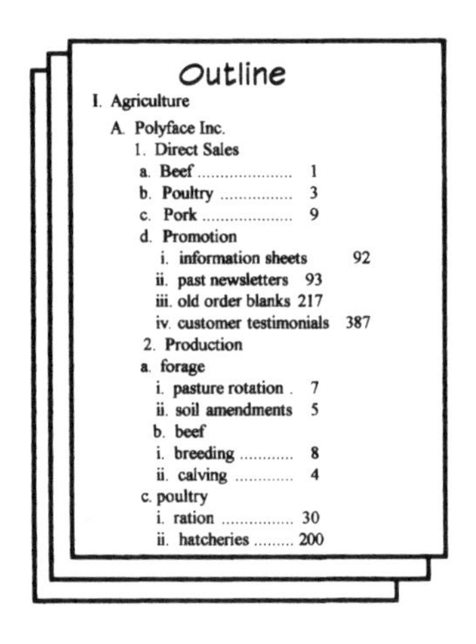
Outline

I. Agriculture
A. Polyface Inc.
1. Direct Sales
a. Beef 1
b. Poultry 3
c. Pork 9
d. Promotion
i. information sheets 92
ii. past newsletters 93
iii. old order blanks 217
iv. customer testimonials 387
2. Production
a. forage
i. pasture rotation . 7
ii. soil amendments 5
b. beef
i. breeding 8
ii. calving 4
c. poultry
i. ration 30
ii. hatcheries 200

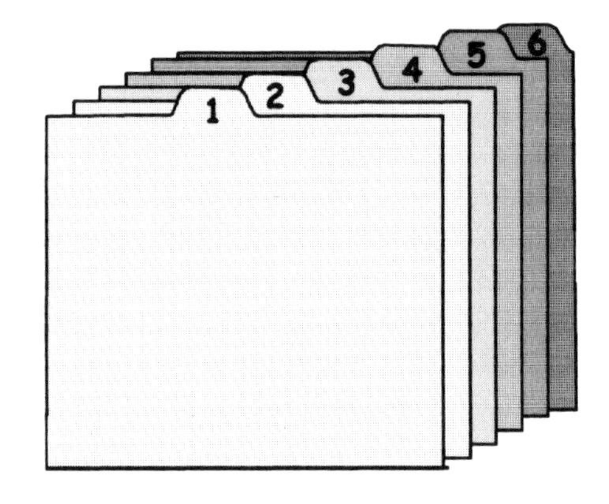

Establishing Your Market

Designer Agriculture

We live in a designer day, a post-industrial info-age when folks are looking for something unique. Societal changes stimulate new awareness, which translates into new opportunities.

I am fascinated by the things I'm reading as we come to the confluence of both a millenium and a century. My eclectic appetite for information has brought me across all sorts of ideas lately. While I do not believe everything I read, the important thing is to understand that other folks are reading these things and these writings reflect the dominant perceptions of our age.

For example, people who divide history into 50-year, 100-year and 500-year cycles point out that all of these are coming together right now. The Birth of Christ, the end of the Roman Empire, the Dark Ages, and the Reformation all happened in rough 500-year increments. And just as the printing press, the invention of gunpowder and reformation revolutionized our world, so are the computer, Promise Keepers and post-industrial employees searching for meaning fueling dramatic changes in our day.

As the information age sweeps upon us, we are seeing a complete breakdown of trust in the old industrial paradigms. Our most sacred institutions are crumbling, downsizing and trying to maintain some vestige of public trust.

What we are seeing is the maturing and necessary demise of the industrial model. Henry Ford lamented that what he disliked most about his factory was that he had to hire a whole man when all he needed were two hands. Sounds like quite a noble ministry, huh? People eventually rebel, either overtly or covertly through creativity, against models predicated on this kind of thinking. A model can only be anti-human for a certain length of time before natural resiliency pushes back.

The fact that gunpowder in a mere decade made a centuries-old institution like chivalry and knights obsolete lends credence to the notion that today's models are no less vulnerable to megapolitical changes. The demise of Roman Catholic Church supremacy occurred not so much because of Martin Luther, but because the church abused its power and the printing press enabled ideas to be widely disseminated. In just a few decades, the church lost control in a worldwide megapolitical breakup. This rising and falling cycle occurs religiously, politically and culturally. All mainline denominational churches in the U.S. today are seeing falling memberships. James Dale Davidson in his recent book, *The Sovereign Individual,* explores this issue in great detail and prophesies, based on historical contexts, where our world is going.

Today cottage industries are springing up everywhere. Outsourcing is the buzzword of the large corporation. Even the Internal Revenue Service is talking to its agents about working out of their home offices in order to reduce town-office rental costs.

The Internet and the universal lack of respect for our large institutions are creating huge opportunities for *designer anything.* We have designer jeans, designer sound systems and designer trail mix. The designer idea merely reflects distinctiveness and fragmentation.

Society-wide dissatisfaction with industrial chicken factories, swine cities, cattle feedlots, *Salmonella, E. coli* and government inspectors is real and spreading. Due to info-mania, people are more aware of government-business collusion and the cozy relationship among multi-national corporations. Keeping evils hidden is more difficult when anybody with a computer and modem can broadcast information to the world from a home office.

For the designer farmer, therefore, the opportunity looms large. This is not the time to do what the mainline institutions do, because they are crumbling. Industrialization gave us what it could and is now in decline.

All of us are somewhat provincial. We tend to encircle ourselves with like-minded folks and to be completely unaware of things outside our immediate world. For example, millions of people actually know what *omega-3 and omega-6* fatty acids are: they are cholesterol reducers. The reason is that several diet and health magazines have subscribers numbering in the hundreds of thousands. Just three decades ago people in our society did not subscribe to even one of these publications, let alone a dozen. None of us had heard of a fax or a cell phone. None of us knew someone who wore a pager and we really believed a job with AT&T was set for life. How society has changed in just 30 years!

The broad-scale disillusionment with existing institutions and the hunger for uniqueness creates an unprecedented opportunity for the with-it farmer. People actually do care whether their pork is from a pig that rooted in the dirt. They actually do care if the farmer spends his day wearing an oxygen mask because he's covering his tomatoes with Sevin. They are concerned about honeybees and butterflies. Not everyone, to be sure, but many, many folks. They are tired of the same-old, same-old and looking for something new. That is why recreation and entertainment are growing at unprecedented levels. Life boredom now permeates our society.

Brand new attachments to the prefix *agri-* lend proof to this basic societal change. We now have agritourism, agrirecreation, agrieducation. Some farmers now earn more money in one month on a pumpkin patch with related things like crafts, concession goodies and putt-putt golf or straw bale mazes than they ever dreamed of earning on 1,000 acres of soybeans or corn.

While the old guard is still pooh-poohing organics because it can't feed the world, many of us are recognizing the huge marketing niche for designer chicken, designer lettuce and designer beef. While the academics sit around and debate how we're going to feed the world, thousands of farmers go out of business because the old institutions representing old markets won't pay the farmer's bills.

I suggest that farmers tap into this growing market with designer products and ride this sweeping cultural change into success. It will mean a fundamental change in how and what we produce, in how we process and how and to whom we market, but the rewards are infinitely superior to anything offered today by the industrial model. Most aging farmers will sit around sipping their coffee complaining about the multinational corporations and the grain cartel, all the while wearing a cap emblazoned with the graphics of the agribusiness entity they patronize.

Many of us are opting out of that scene. We're not complaining; we're changing and doing, accessing the new opportunities created by our changing culture. People hunger for a relationship to food, to land, and to people who care about those things. They want better food. Recycling is only a phenomenon representing a broad megapolitical change in our culture. Designer farmers are already living in the future. Will you join us?

The key to developing a designer product is superiority. You cannot pull people away from conventional markets unless you offer them something superior. What people want is integrity and character in their food. Establishing this superiority requires several things:

HANDLING AND TASTE. No matter how much hype you put on the product, if it doesn't handle well in cooking or doesn't taste good, you can't sell it. It absolutely must be appealing in these areas.

When we first went to restaurants with our pastured eggs, we called and made appointments with five chefs. By the way, if you want a premium price for your food, concentrate on restaurants that have chefs, not cooks. We scheduled appointments one hour apart.

I just called these chefs cold turkey and said: "Hello, I'm Joel Salatin from Polyface farm and we produce the world's best egg from hens on pasture. Could I come by and show it to you?"

Every single chef was glad to look at the eggs. We took a sample dozen for each chef. In every case, the restaurants were purchasing commercial eggs for 60 cents a dozen. Our price was $2 per dozen. We had no brochure, no info-packet; we didn't even have a label. These chefs did not know me from Adam.

Anyone who has been in sales knows that we were foolish beyond belief to do this. You don't go in cold turkey with a product and ask the buyer to go through the hassle of changing vendors, plus pay 3.5 times the price! This is suicide. But nobody told us it couldn't be done, so we just waded on in.

We went to the first restaurant and the chef took one of our eggs, cracked it and dropped it into a saucepan of almost boiling water. As soon as the egg hit the water, he began to get excited.

"Wow! Look at that!" he exclaimed. Then he pulled the egg out with a slotted spoon and placed it reverently on a saucer. He began to rub it with his fingers, kind of massaging it gently. I was beginning to think this was a little weird, but I'd never made a cold sales call on a chef, so I figured I needed to learn too.

By this time he was downright ecstatic. I couldn't figure out what the big deal was. I asked him: "How does this egg differ from the ones you're used to?"

He didn't even answer. All he did was reach over into a flat of his normal eggs and toss one into the same saucepan. As soon as it hit the water, it virtually exploded. The yolk came apart from the white and the white spiraled off into a thousand convoluted ringlets. Then he tried to pick it up with his slotted spoon and got about half the pieces.

Needless to say, I was dumbfounded. As it turned out, this restaurant had a specialty Sunday brunch centered around a hot water poached egg. The chefs had just been discussing discontinuing this centerpiece menu item because they simply could not get the eggs to hold together.

Needless to say, he went ballistic over our eggs and bought 30 dozen on the spot. That restaurant has remained faithful now for years, and has increased to 60 dozen a week.

At the next restaurant, which is the biggest one we service, I cracked out a couple of eggs on a saucer and immediately several chefs gathered around in amazement. They exclaimed about the color and how high it stood up on the saucer. Then one of them reached over and pulled the yolk up with his fingers. The white came right up with it and the whole group, in unison, said: "Look at that. I've never seen an egg do that."

They were excited and told me to go across the street to the bakery and show the pastry chef. I walked over there and cracked out an egg for him. Then I pulled up the yolk and watched his eyes widen as the white came right up with it. Then I threw the yolk back and forth in my hands as I talked to him, and he was spellbound.

He told me that he had such little success being able to consistently separate the white and yolk that he had almost quit making things that required a separation. This ability to consistently separate the yolk was truly amazing.

A few weeks later, the invoices for these eggs finally went across the comptroller's desk and he hit the ceiling. "What? What are we doing paying $2 a dozen for eggs?" He immediately demanded the chefs to come to his office.

I wasn't there, so all I can report is what the chefs told me, but they told him about the realities of the kitchen: "The old eggs we were using always had one broken one in a flat. That one leaked over on its neighbors. Trying to get those eggs, which were stuck to the cardboard, out, we always end up breaking a couple more. Then when we finally crack an egg onto the griddle, half the time the yolk breaks. We have to scrape that egg off into the garbage and start over."

Then, collectively, the chefs said: "Mr. Comptroller, we can make more money in our kitchen on $2 a dozen eggs than we can on 60 cent eggs."

Another pastry chef makes a 6-inch cake. I don't know how much they sell for, but knowing the restaurant I'll guess $20. Her recipe makes four cakes. When she switched to our egg, without any recipe change whatsoever, the batch made five cakes. She made more money with the expensive eggs than with the cheap eggs.

In addition, she found that using these eggs reduced spoilage. The normal window of marketability for pastries is 1-2 days, but she found that by using our eggs she could extend that to 3-4. That may not sound like much, but doubling the window of marketability makes a huge difference in spoilage.

One of our chefs changed restaurants and of course took us as a main supplier with her, introducing our food to this new place. The first time she got broilers, the staff worked them up in preparation for the dinner hour and began commenting: "Our hands aren't sore. Why is that?"

The "sore hands syndrome" is common in food establishments that handle poultry. The juices have enough manure and chemicals in them that after a couple of hours working with chicken carcasses these penetrate your hands, producing a sore, hurting reaction. It is an accepted norm in the food processing industry. But with our birds, even after working with them for hours, no soreness develops.

"I can feed twice as many people with a pound of your ground beef than I can with what I get out of the supermarket," said one patron after she began buying our beef. The reason? Ours is real meat and it is lean. Pasturing rather than grain feeding produces a different type of meat.

Down at the farmers' market we've taken beakers of the broth drained off a pound of fat from lean supermarket ground beef and a pound of ours. Not only is the volume dramatically different, but so is the consistency and appearance. The supermarket fare doesn't look edible and gets real hard, indicating saturated fat.

The same is true with poultry. We cook identically sized broilers in identical slow cookers and drain off the broth to illustrate the difference. The broth from the supermarket bird is actually gray — indicative no doubt of the fecal soup. Ours looks like amber honey.

I could go on and on with these stories, but I think you get the picture. The same is true with high quality organically grown fruits and vegetables. Sweeter carrots, juicier apples, tastier cabbage, more tender tomatoes — all these elements make for superior taste and handling. Creating distinctive superiority starts on your farm.

EDUCATION. People need to be reminded over and over again about the difference between your food and the conventional stuff. Use phrases that educate but also play on the emotions.

A friend of mine gave me the following tag line for our poultry: "Our chickens don't do drugs." That's an exceptional slogan. It takes a phrase in common usage, with attached emotion, and uses it in a catchy way.

We made a sign for our beef after mad cow disease took the headlines:

"OUR COWS DO NOT EAT DEAD COWS OR CHICKEN MANURE"

That sign sure raised some eyebrows. The average person has no idea that it is standard in beef operations to feed chicken manure. Cow colleges have stacks and stacks of research projects showing that cattle eat it all right and the beef tastes okay and we should all do it. But I have not run into anyone yet, on the street, who agrees with the "scientists." Folks classify these USDA eggheads as pawns of the poultry industry and quit buying supermarket beef.

Mad cow disease and all these seminars on feeding chicken manure to cattle are our best advertisements. People look for options in droves, and we're happy to oblige their quest. Do not overstate the case, but be bold. I call supermarket eggs "fecal eggs." If you use this type of phraseology matter-of-factly, without a lot of hoopla, people respect it instead of thinking you're a kook.

"Concentration camps" and "factory farms" are good phrases to get the point across. Perhaps some would accuse me of being too radical, but my experience is that using these phrases grabs people's attention in an inoffensive way. People consistently come up and ask: "Do cows really eat chicken manure?"

I am always amazed at people's provincialism. Since I see the agriculture press and read all the articles extolling the virtues of feeding chicken manure to cattle, I assume it's common knowledge. But it's not. The average city person doesn't read those articles and has no memory of the cursory attention paid to the issue. It simply does not penetrate the urbanite's life in a meaningful way.

I've had customers at the farmers' market come back to our booth and say: "We just wanted to stop back by and tell you how much we appreciate you educating us."

A knowledgeable customer is a discerning customer, and a discerning customer is a loyal customer. Whenever a publication comes out with an article about food safety and the filth in large processing plants, we display it prominently in our sales area. We quote succinct press reports about problems in the industrial food system to keep our people aware of what is going on.

Part of our education is to encourage people to vote with their food dollars. Our culture places way too much emphasis on dealing with problems by going through the legislature. Rather, we would like for millions and millions of people to opt out of supermarket fare, voting every day in the marketplace with their dollars. When the mainline systems begin feeling the pinch in the pocketbook, perhaps we can get their attention. If we wait until 51 percent of the people wake up and pass legislation, it will probably be too late.

Taking the education one step further into proactivism by patronizing your clean, green, humane product is the ultimate customer loyalty. When your customer believes a purchase from you not only fills the dinner plate with better food, but also moves the culture toward truth and righteousness, you have a loyal customer.

STORYTELLING. Use descriptions that play on people's idealism. Folks generally have a conception about what a farm is supposed to be like. Potential customers think animals should move out on pasture. The humane issue is real; the Humane Society in the U.S. has about 1.8 million dues-paying members. That is huge. Here are some words that excite people and make them want your product:

- Humane
- Local
- Fresh
- Ecological or environmental
- Clean
- Drug-free

The idea of local production has a warm place in almost everyone's heart. Keeping dollars in the neighborhood, patronizing local producers, and knowing your farmer are all things that give your customer an emotional attachment to your food. It's everything that a label isn't, and that's what you want to emphasize.

Play on the personal relationship angle. "Do you know where your food comes from?" Take folks on a farm tour and open doors for them, announcing grandly: "No secrets here." People respond to honesty and openness; this is what they know they are not getting from the USDA, FDA and every other government agency, right up to the White House. They know the multinational corporations don't tell the truth and the vertically integrated food industries don't tell the truth.

We've always told people: "Anyone who thinks we're using chemical fertilizer, pesticides or medications is welcome to come unannounced any time of the day or night to check us out." Of course, no one has ever taken us up on it; just saying it dispels all fears that we're trying to pull a fast one.

What people desperately want today is a link to the land, a nostalgic story to tell about their food. Position yourself as a family enterprise where vegetables get tender loving care. Where children run through the fields and pet the animals. Where soil microorganisms receive steady doses of compost and long nights of rest, protected from harsh chemicals and unmanaged, marauding livestock.

What you want to create is the perception that your customers have a stake in your farm. In fact, the ultimate is to get them to call it "our farm."

A farmer in the Midwest actually capitalized on this idea and named his farm "OUR FARM." It worked like a charm. Whenever his customers were going to come out to buy things and their friends would ask them where they were going, they would reply: "Our farm."

When they would go to a social gathering and report on what they'd been doing, their children would tell the other children: "We went out to 'our farm' this week."

While this may be unnerving to some farmers concerned about property rights, the benefits outweigh potential problems. Besides, these folks are not interested in taking over; they like what you're doing and simply want to identify with it. To offer them that degree of intimacy with what you're doing will protect you from lawsuits, sustain your customer base through lean years, and encourage you emotionally like nothing else.

Encouraging folks to pet the chicks, pick the strawberries, pull a carrot and picnic under the maple tree will stimulate the positive experiences folks can relate to their friends and tuck into their memories. Stories simply verbalize our memories. Anything you can do to create a better memory, a more vivid memory, a more exciting memory, will help to impress your story on your customers. Your story will become their story. An individualized story not only creates the best advertisement, but also personalizes the identification with your farm.

Creating the superior product is tied in with having a better story to tell. While some people may be awed by the efficiency or sheer size of the industrial food producing model, when it comes to what a person yearns to put on his plate, his soul returns to Grandma's garden and the chicken yard. That's the real story.

And that conjures up tastes, textures, sights, smells and memories beyond anything Jolly Green Giant can put on a can. You cannot fool people with false superiority, although many businesses try. Ultimately, the cream rises to the top. Excellence will always be discovered.

Your marketing niche is directly related to your pursuit of

excellence. Developing unparalleled product superiority enables you to be at peace with yourself, your community and your customers. It is the first step in marketing. Go for it.

Developing Your Farm's Clientele

"How do you get customers?" That is perhaps the most common question I am asked. It is certainly fundamental to any marketing effort. Really, the question boils down to a more basic, simpler one: "How do you get your first customer?"

We practice what I call the 3 'Es': **EDUCATION, EXAMPLES and EVANGELISM.** Let's take them one at a time.

EDUCATION. We put together an educational slide program about our alternative farming methods. It showed how we healed the land with livestock and how we encouraged wildlife by diversifying the landscape. We have slides of composting, gardening and pastured chickens.

Every community has organizations that meet periodically and want some sort of after-dinner educational program. In addition, many assembly-type opportunities exist in schools. Here is a listing of some of the local organizations or groups to whom I've presented this program:

- Kiwanis Club
- Ruritan Club

- Rotary Club
- Women's Club
- Junior Women's Club
- Soroptomist Club
- Garden Club
- Key Club
- Exchange Club
- American Association of Retired Persons (AARP)
- Retired Teachers' Association
- FFA Chapters
- 4-H Clubs
- Women's Circle groups in churches
- Young Farmer's groups
- Science Clubs
- environmental studies in school science classes — grade-wide assembly
- Leaders 2000
- Toastmasters
- Natural Resources Conservation Service banquets or annual meetings
- environmental groups
- Isaac Walton League

Many of these local groups have invited me to speak several times. What happens is that the membership changes and a new program chairperson comes in who doesn't realize that I was there two years ago. Of course, since we're always refining things, the program is never the same.

All these outfits are starving for interesting programs. They get tired of listening to the rescue squad fundraising chairman, the United Way Chairman, the school superintendent and all about the latest homeless program. They want something new, something different.

Every evening on the 6 o'clock news it seems there is more bad news about how agriculture is destroying the earth. People see filthy processing plants cranking out manure-laden poultry and beef. They see space-suit-clad farmers applying herbicides and pesticides to huge fields of head lettuce or celery. They see the pall of fecal dust hovering over feedlots, Mexico's sewage-irrigated strawberries being served in school cafeterias, and concentration camp veal calves. They read reports of swine manure spills and mad cow disease. In all, they receive an education steeped in anti-farmer sentiment. Farmers bear the brunt of finger-pointing and blame.

But people enjoy eating. They want to eat hamburgers without feeling guilty about destroying the earth. They enjoy biting into a tender pork roast, but hate the thought that it's full of medications and probably contributed to the swine facility's manure spill that killed a million fish last week. People are emotionally torn between what they want — indeed, what 85 percent of us need — and what they understand to be socially and environmentally responsible buying habits.

And here is where our slide program comes in. The notion that we can use livestock to upgrade the environment, that we can produce healthy animal proteins without vaccines and medications, hits consumers like dynamite. They've been looking for healthy food. They've been yearning deep in their souls for hamburgers that don't destroy the earth. And suddenly they see how it can be done.

We show them how we mimic nature in building soil, growing plants and managing livestock. It's like a breath of fresh air when they realize no farm needs to stink up the neighborhood, that perennial polycultures with high density, short duration grazing can mimic climax prairies of old, and that forage-based beef can be tender and nutritious. Boy howdy! They go bananas!

The key is to *not* give a sales pitch, but to concentrate on education. Concentrate on the following issues (quoted from our Polyface promotional brochure):

- More nutritious. Lower fat and higher vitamin/mineral contents. You can taste and see the difference.
- More humane. Animals receive fresh air, sunshine, exercise and species-correct grouping on fresh green grass. Pigs

can root and wallow. What a life!

- More neighbor-friendly. Proper modeling and management ensure only pleasant odors, pleasant sights, and pleasant sounds. Good farming appeals to all the senses.
- More customer-friendly. Patrons enjoy a relationship with the farmer and a link to the land. We build bridges instead of barriers, friendships instead of foolishness.
- More family-friendly. Children enjoy seeing and petting the animals, romping in the fields, and learning firsthand who, what, where, when, why and how about *their food.*
- More tasty and beautiful. The proof is in the eating. Adorn your palate as well as your plate with the rich color and taste of our products.
- More environmentally friendly. Patronize an agriculture that heals and regenerates, that thrives on diversity, resource regeneration and lower petroleum use.
- Chemical free. Our animals thrive without antibiotics, hormones, and systemic biocides. Healthy food for healthy people.

In addition, here are other themes that may permeate a presentation:

- Preservation of rural viewscapes by reducing the need for farmers to sell out in order to make any money. Profitable farming reduces sales pressure.
- Local economy. Keeping dollars turning in the community reduces their export to far-flung places.
- Reverence for biology. Plants and animals are not just machines. Assaulting creation with a mechanistic, industrial view is not only immoral, but also devastating in terms of environmental degradation, people's quality of life and quality of food.
- You can be part of the solution. This is key. The realization that people can actually encourage a clean agriculture empowers them to take action.

At this point the audience is ready to act and I show them a couple slides of our customers at the farm purchasing clean food. Suddenly they understand that this is not just some theoretical thing, it is something in which they can participate. Now I give the altar call: "I know some of you bitterly oppose concentration camp farming and chemical agriculture. You join organizations like Nature Conservancy, Sierra Club or the Audubon Society. You know how smelly chicken factories are. So today I've set before you a choice. You can choose to aid and abet the industrial, chemical, pharmaceutical approach, or you can encourage local clean food producers. It's your choice.

"What kind of society do you want for your children and grandchildren? What kind of community do you want to live in? Look, if you really dislike factory farming and the environmental degradation it causes, isn't it reasonable not to patronize those food establishments that are the product of such abusive practices? A whole world of clean food and clean farming exists in direct opposition to the factory mentality. Which side will you choose?"

This is what you call "putting it to them," but you almost have to hit people over the head with a two-by-four in order to get their attention. You may feel like I've done that some in this book. Well, I'd rather be a bit offensive and have you get the message than be wishy-washy and have you continue to be the same person you've always been. A passionate message changes people.

Of course, I always end the presentation with an invitation to everyone to come on out and visit the farm, to see firsthand if these things are true. The sales pitch occupies such a tiny element that no one has ever accused me of using this presentation as a sales pitch. I would say most people do not even realize they've heard a sales pitch.

Depending on the group, I'll do different things to augment the slide program. For example, if it's a science class, I'll give each student a Polyface brochure to take home to their parents. Obviously I custom tailor each presentation to fit the group I'm addressing.

Garden clubs, which generally meet in a hostess' home, are particularly enjoyable. Not only is the setting more informal, but women consistently respond to my program more than do men. Women are the ones who take the children to the doctor, the ones who sit up with the children at night. Women have a more intuitive understanding that "this is right." Men sit there all macho and demand empirical proof, the old "show me" thing. Often, men just don't get it.

At garden club meetings, then, I'll take a dozen eggs and ask the hostess for an egg from her refrigerator (assuming she's not a customer) and two saucers. I break her egg out and then break one of mine. The differences are obvious and the ladies realize that indeed there is a difference. That's the critical thing. Then I give the rest of my dozen to the hostess as a hostess gift. My cooler with 30 dozen in it is sitting by the door. As soon as the meeting is over, they drop their money in a box and take eggs — strictly honor system because I'm busy visiting.

Another favorite garden club technique is to take a slow cooker with a beef roast or steak cooking in it. I plug it in when I get to the house where the meeting will be to keep the roast warm while I talk. When I'm done I take a meat platter and pull out the beef. When we do this, we put in no seasoning and no water. Of course, they carry on about how it just falls apart. Then I break it up and have a canister of toothpicks for everyone to try it.

At all these presentations, I have order blanks available where folks can pick them up and become customers right on the spot. Generally each of these meetings will bring in 2-4 new customers. Most will need to be touched several times before they see the light. That's okay. Non-customers do not oppose you; they're just customers-to-be. Give them time. They might be just one television news program away.

AARP groups always give out door prizes during their meetings. I think they do that in order to keep the members from heading home for a nap after the dinner. Anyway, I usually take a couple of chickens or a few dozen eggs and add them to the door prizes. These groups love it. It shows that I can get in the festive mood and the officers enjoy being able to give out another half a dozen prizes. Of course, for those who didn't get a prize, I have my cooler of eggs for sale by the back door.

What you can do is only limited by your imagination. Your goal is to get across two points:

(1) Food quality varies widely.
(2) Your food is better.

Let me touch on one other point before we go onto the next 'E' of marketing. I find that many folks have a stereotypical view of organic farmers as kind of "look-what-the-dog-drug-in" hippies. By going directly to consumers, I allay these fears, breaking down those preconceptions without ever uttering a word. The mental picture of the farmer redneck, too, is important to break through. When they see an honest-looking fellow giving a passionate talk that is educational and articulate, they come around real fast.

This is why I think it is necessary to advertise an unconventional product unconventionally. To be able to stand right in front of folks, open yourself up for questions or argument, and defend your position will get you farther quicker than thousands of dollars' worth of normal advertising. Although it does not reach as many people, the people that are reached have a clear understanding of you and your product.

EXAMPLES. Whenever a company introduces a new product, one of the first things it does is give out samples. Proctor & Gamble puts shampoo samples in your mailbox for a good reason – you are resistant to trying something new.

People just don't walk down the aisles of the supermarket saying: "You know, I've been using 'X' brand shampoo for ten years and today I just have a hankering for something different." We just don't do that. Manufacturers know that to get you to change from whatever you've been using to something new, giving out free introductory samples is one of the most cost-effective marketing tools available.

Permutations on this theme are everywhere. Free introductory magazine issues and test driving a car both work on this "try it at no risk" idea. The beauty of a sample is that you are always more emotionally predisposed to like a gift than you are something you paid for. Paying for something heightens your critical judgment, whereas a gift heightens your forgiveness.

When we first started, we gave T-bone samples to folks. We fought the prejudice against grass-fattened beef ("don't you know it has to be corn-fed?"). The only way to overcome that prejudice was to get the meat into peoples' mouths. It works.

In its most fundamental form, marketing is simply taking something you produced or made, walking over to the neighbor's house, knocking on the door and saying: "I'd like you to try the world's best chicken, or whatever. Here is a sample. I'll be back in touch to see how you liked it."

That really is all it takes. The first one is always the hardest. Each subsequent one gets easier because you develop courage and finesse in your approach. But if you believe that you have the best, telling other people about it is easy.

Develop a hit list. I know, I know, that's not a good term — marketing specialists would call this a prospect list. Brainstorm onto that list any potential customer. You can list name, establishment, whatever. Then go back and prioritize from most likely to buy to least likely. Often this list will start off with friends and relatives, or coworkers and people with whom you do business. Here is a sample:

- Mom and Dad
- Brother and sister
- In-laws
- people in church
- relatives
- neighbors
- co-workers
- doctor
- lawyer
- hairdresser
- auto mechanic
- banker
- dentist
- plumber
- electrician
- herbalist
- carpenter
- car salesman
- insurance salesman
- homeopaths

- naturopaths
- wellness center
- fitness center
- lawnmower mechanic
- gourmet restaurant
- children's school teachers
- school principal
- congressman
- senator
- governor
- president
- meter reader
- bank teller
- accountant
- local hospital dietitian
- vacuum cleaner salesman
- local politicians
- local bureaucrats
- minister
- guru
- any business associate
- veterinarian
- filling station owner or attendant
- people in civic club or other charitable organization
- telephone solicitor — turn the pitch around for a change
- any sports affiliate — coach of children's soccer team; all the players' families
- multi-level distributor of anything (Amway, Mary Kay, Avon, Shaklee, etc.)
- anyone else you can think of

Over the years, we have given samples to lots of people, but it seems like whenever we give something away, the return is fourfold. Many farmers do not like to give things away to city people because they think city people have been giving them the short end of the stick for a long time. Believe me, it's not the people; it's the system.

If you are the first one to give things away, others will respond accordingly. We use this at the farmers' market. The first year we went down there we had a huge surplus of pullet eggs because of a glitch in our marketing effort. We have given lots of eggs to the local homeless shelter over the years. But even they can only use about 30 dozen a week. We had 600 dozen!

For the opening two Saturdays of the season, we just gave away eggs. We handed out 600 dozen pullet eggs. People loved it. You should have seen their faces when we gave them free eggs. A couple of the other vendors who sold eggs were not too fired up about it, but it sure brought people down to the market. The increased number of customers to the market more than made up for whatever losses those vendors may have suffered.

"Welcome to the farmers' market. Have a dozen eggs," we said as people walked through the market area. Once we got through that two weeks, we quit giving away eggs and began selling them for 50 percent more than anyone else at the market. But we had a firm toehold. People had tried our eggs and were sold on them.

We give away freebies as come-ons to purchase more. This is the way we like to use overages. No matter what you do, there will be times where you just have to salvage things. I would rather get marketing mileage out of a surplus item than drop the price.

If you can use both the surplus and the free introductory offer at the same time, you get more mileage for the gift. For example, when we began the commercial-sized pastured egg enterprise, we had hundreds of spent hens to sell.

Most people do not know what a stewing hen is anymore because in the industry these birds are inedible by the time they've laid for 10 months. Healthy, heavy stewing hens basically are not available in the marketplace anymore.

In order to introduce them, we began giving them away to anyone who made a $20 purchase or greater. Of course, we'd tell folks to be sure and slow-cook or pressure-cook these birds. The taste is out of this world, but older hens do have to be cooked longer. Within a couple of weeks, we had people coming by to buy these birds.

The first year, here at the farm, we gave a stewing hen away to many of our broiler customers. They were tickled to get something free, we got wonderful public relations out of it, and they were introduced to a new product. The first year orders were only about 50. The second year it went to 150 and by the third year, we were up to 600. The samples created the market and increased sales and goodwill at the same time.

If you are in the vegetable or fruit business, you can move some less-than-premium product this way. If it stimulates additional sales of your premium stuff, often you will make more than by dropping the price on the stale stuff low enough to get rid of it all.

I cannot tell you how much fun it is to give things away. I would say on average, at least every one of our customers has been given something at least once. This is an ideal way to handle inventory adjustments. If you can take your overruns and turn them into additional customers or twice as much sales at your premium price, you will never regret the write-off.

The bottom line is that examples work. They always have and they always will.

EVANGELISM. Let's say you went down to the local hardware store this afternoon and the proprietor met you at the door saying: "Good afternoon, Jim. You're such a great customer I'd like you to just go over to that tool board and take your pick of anything on there. It's on the house today."

After a moment of stunned retrospection, you would walk over to the tool board, choose an item, and then literally turn into a charismatic. You'd go hip-hopping down the sidewalk telling people: "Hey, you need to go over to XYZ Hardware. That's the greatest place in the world!"

Over the years we have tried to turn our customers into evangelists. The way to do that is to make the customers feel needed and appreciated. Whenever we receive a phone call from a new customer — or cheerleader, we call them — our first question is: "Where did you hear about us?"

Sometimes the person doesn't know, but usually we can trace it to a specific customer. We make a notation on that customer's card in the file so the next time she comes out to the farm, we give her a free something. If she's a chicken customer we give her some beef.

If she's a chicken and beef customer, we give her pork. If she already gets chicken, beef and pork we give her eggs. We like to take advantage of these opportunities to spread the awareness to other products, you see. A little pat on the back, a little appreciation really turns people on.

Our society is starved for appreciation. I know when I get

fired up about a product and tell other people about it I always wish that its producer could know I'm sending people their way. Not that I'm looking for a gift, but just because I believe in the product and want to communicate that to the business, to let the folks there express their gratitude for my patronage and loyalty. Many times no mechanism exists to reciprocate the favor.

We go out of our way to let our folks know that we appreciate them spreading the word. This may not be the fastest advertising in the world, as far as quick growth is concerned, but it certainly brings you the most loyal customers.

One distinct advantage of having our current customers bring in more than 50 percent of our new customers is that the new generation comes already screened.

Have you ever run a cash register or clerked in a retail store? You know how there were always certain customers that nobody wanted to help? And the manager or owner always told you about making sure customers X and Y and Z paid with cash because their checks bounced?

The beauty of screening your future patrons through the discriminatory screen of your current patrons is that you don't have to deal with the riff-raff. When you advertise through the newspaper or other conventional media, you have to go through this screening process yourself. When the public responds to your business, you have no way of knowing if this person is litigious, disagreeable, or writes bad checks.

Having our team players pick the next generation of team members avoids all the business inefficiency of going through tryouts. These new folks will be most like the good old team.

I had a customer one time tell me she had had some neighbors over for dinner and served our chicken. The guests just carried on about that chicken. During the visit, our customer had some differences with these neighbors. Later, as they were leaving, the neighbors asked where she got that delicious chicken. Our customer did not tell them. She told me: "Something about them made me feel like they would not have been good customers for you, so I just didn't tell them where I got the chicken."

How about that? No insurance policy, no application can protect you better. The kinds of people making good customers now are the ones who will bring you good customers tomorrow. By turning them into evangelists, you can encourage what every business covets – word-of-mouth advertising.

Getting customers starts with one. Don't worry about a hundred until you have one. Obviously using patron contacts to magnify this evangelistic outreach is wonderful. For example, if you have a customer who is a chiropractor, he may be willing to add your farm brochure to the freebie reading rack. The same could be true in the wellness center or the local fitness center. Plenty of health-minded places, especially if those run by your customers, will gladly place your material on a reading rack or bulletin board.

Capitalizing on your loyal patrons, with all their energy and their contacts, will grow your business steadily and efficiently. As you meet the needs of people they will respond beyond your wildest dreams. That is what marketing is all about.

Communication

At the heart of nearly every successful business endeavor is a gifted communicator. To enlist support, convince buyers and motivate partners requires speaking and writing skills.

While this chapter certainly cannot replace whole books on creative writing and persuasive speech, it may give some basics to help you in this area. Our society is good at producing technicians. But just sit back and think about how many speeches or lectures you've ever heard that were spellbinding and you will realize what a rarity good oratorical skills really are.

Just to substantiate the importance of this topic, here is a list of oral presentations you may need to make:

- Convince a farmer to let you help him
- Ask a church or fellow club member to let you use a few acres
- Persuade a relative or friend to give or loan you some capital
- Explain why your tomatoes, chickens or flowers are the best in the world
- Negotiate the price of a used tractor
- Speak to a civic club about your farm and why they should patronize your products

- Teach a group of elementary school students about your farming methods powerfully enough that they will go home and tell Mom and Dad to buy your products
- Tell a bureaucrat to take a hike
- Warn a neighbor to keep his cows on his side of the fence

Here are some writing jobs you may need to do:
- Directions to your farm
- A brochure for publicity
- Patron newsletters and order blanks
- Letters to the editor
- Requests for clarification or information
- Magazine articles
- Guest columns in the agriculture section of the local newspaper

Writing and speaking are like anything that requires skill: the more practice and experience you devote to it, the better you will become. While it is true that not everyone will become a best-selling author, almost everyone can develop enough competency to feel comfortable communicating. Putting time and energy on it will do wonders for your ability and confidence, which in turn can begin to take the dread out of it.

Only a truly gifted person becomes a Paul Harvey or a Dr. Martin Luther King, Jr., but plenty of people become tremendous speakers. I think the most important point about speaking and writing is that you will never become good at what you refuse to practice. My goal here is to give you enough ideas to push you over into doing that first presentation or that first newsletter. Getting off dead center —or "dread" center — is the hard part.

Get excited. The best communication is always from passionate experience. What moves people is the force generated from your own movement. If you are not "fired up" about something, you cannot move anyone.

This level of hype is something you alone can generate, but it is absolutely necessary if you are ever going to communicate effectively. Society is full of drab, blah people who simply go where the current takes them. They have a dead fish handshake and a deadpan face — just look at the hollow faces of kids at the mall and you can see it.

To have an animated countenance, a spring to your step and a zest for sharing your nuggets of truth is not *a* but *the* prerequisite for effecting speaking and writing. If you're on fire, the world will come to see you burn. Some of the most moving presentations I've ever heard have not used good grammar or fluent articulation; they come from the heart. Yes, style is great, but a "fire in the belly" is far more important.

So how do you get this excitement? First, I think you need to become educated about your topic. I'm not talking about institutional academic education. I mean reading books and articles, visiting knowledgeable people and picking their brains, attending conferences and dedicating yourself to being an expert in your field.

Only 30 percent of Americans ever read a book cover to cover following high school graduation. That is a phenomenal statistic. It's devastating to folks who write books, and it's a sad commentary on our educational system. But what it means is that very few people are willing to stoke any kind of inner boiler at all.

A friend of mine pointed out that to be successful you don't need to be that much better than everyone else is, because everyone else is so mediocre. If you begin reading just one book per year, you'll be in the top 25 percent of our culture by this measure. If you begin reading a book a month, a much higher group. If you lack book-buying money, use the local library. Ask the librarian to order books that interest you; this is a favor to the library and other like-minded folks.

Instead of going on Disney vacations, visit successful farmers across the country. Andy Lee and Pat Foreman, authors of *Chicken Tractor* and *Backyard Market Gardening* say that their favorite pastime is to visit good vegetable operations. That is the way to get excited, and it shows clearly in their writing and speaking. Throw the television out the back door and build a bookshelf in its place.

Attend conferences where you can hear and meet good farmers.

Put attention on feeding your passion. Don't spend time with the complaining farmers down at the coffee shop. Don't spit with the cattle buyers at the stockyard. All they will do is take the wind out of your sails. Hobnob with folks who have like passions and you will soon begin to get excited. This is why Amway distributors attend meetings all the time.

As your knowledge level increases, you will be able to know why your way is the best way and have more confidence to articulate it. As you fill up on the inside, it will naturally overflow, and this overflowing will be excitement. It will also be effective communication.

Get realistic. You will not convince everyone. Do not take a negative response personally. No matter what you do, many, many folks will never agree with you. That's fine. It's their privilege. The greatest communicator of all time was crucified — don't ever forget that.

Many people hate to begin taking a message anywhere because they fear being rebuffed. Look, that's just part of life. Your message is not to those who will not hear, but to those who will. Thank the ones who tell you early they're not interested. That just lets you leave them sooner and go on to the next person who may be interested.

Remember that everyone is on a different point on the line when it comes to persuasion. If you are a 10 and you're talking to a 1, for example, your goal is not to move the 1 to a 10, but to move the 1 to a 2. If you do that, you are successful. If you attempt to move a 1 to a 10 too quickly, you will alienate him. The world is full of people trying to persuade too fast.

Just be content to move people one little step at a time. Many a preacher burned out because he set out to set the world on fire and found that the best he could get was a slow burn. Be glad for the slow burn. In your writing and speaking, keep reality in perspective and it will keep you from being discouraged. It will also help you focus your attention on the next point.

Be affirmative. Now I know you may be wondering how I can say this inasmuch as I've taken some potshots at the "opposition" in this book, but the overall theme is one of positive alternatives. Keep a smile on your face and always offer a creative answer to the ones you impugn.

Rather than just lambasting confinement factory farms, for example, focus on how good, clean and nutritious pasture-based meat is. Rather than talking down the opposition, build up your side.

This is a classic debate principle. The two sides in a debate are affirmative and negative. Affirmative always takes the side for change; negative takes the side for the *status quo* or what is. As the affirmative side presents its case, it must offer both a case and a plan. The case is the problem in the current system and the plan is a solution.

All the negative side must do to win is to refute either the case or the plan. The affirmative must carry both parts in order to win. For example, if the affirmative loses the case but has a strong plan, or solution, the debate round goes to the negative. The reason is because it doesn't really matter how wonderful a solution may be, the only reason to change the system is if the system is flawed.

If the current system is fine, then it is too risky to change models. By the same token, if the affirmative wins the case and the judge agrees that indeed the current system is fatally flawed, but the negative successfully argues that the affirmative's solution cannot or will not solve these problems, the negative wins. The affirmative burden is both to impugn the current system and to present an effective alternative.

Always remember that being negative is easy — all you have to do is disparage and tear down. Being affirmative is much more difficult, not because it means never saying anything nasty, but because it means providing a workable alternative. When people complain to me about things, I always ask: "What would you do?" You'd be surprised how many people stop right in their tracks when you hold them to that point. They don't have a solution.

That is especially effective with young people. Next time you hear a bunch of grumbling, ask for a positive alternative and subject it to the scrutiny of truth. Plenty of gripers would be put in their place in a hurry if simple classical debate theory were put into practice.

In your communicating, then, for every disparagement quickly offer a positive alternative, so that your theme clearly goes forth as an affirmative, creative message.

Get basic. I think history will record that the single biggest contribution Allan Savory and Holistic Management have made to our culture is to push for a single goal. It provides the glue to hold all the pieces (wholes) together and the vectors to keep them on course through expansion.

The first chapter in any composition textbook deals with developing the thesis statement. Until your message can be condensed into one sentence, you are not ready to communicate it. Any time you get ready to make a presentation or write something, just take a minute and put your main objective into one statement. That is the essence of what you want to communicate.

Have you ever been accused of rambling? Many of us lose our audience because we've lost ourselves. Until we have a clear direction we cannot expect to explain our point to anyone else. Wasting everyone's time with fuzzy messages is something we practice too often. Boiling our point down to a precipitate will keep us on track. It will make us proceed with clarity and purpose.

A corollary of this point is to hang a skeleton down from this thesis. Most of us are familiar with outlining, and although it may have fallen into some disfavor in modern days, it is still a wonderful tool. I use it all the time. Here is the basic outline form (in case you forgot).

I.
  A.
    1.
      a.
      b.
    2.
      a.
      b.
  B.
    1.
    2.

II. (etc.)

What this does is give you organization. Certainly the better you get, the less you need to get down into all the detail, but having your message laid out in this organized way gives clarity to it. Often when I get ready to write a column or letter, I just scribble down the basic points like this to keep from getting out of logical order and to make sure I don't miss something as I go through. Parallel thought and structure add strength to your argument in addition to making it more succinct.

Going through this exercise will help the communication flow, creating a sequential momentum that draws the audience along. Most of the chapters in this book have been laid out in this fashion. Again, while I do not want to appear condescending to anyone by going over some fairly elementary composition mechanics, I also appreciate how easy it is to just take shortcuts or to bite your fingernails and spin your wheels, not getting anything done because you don't know where to start. These are starting points and will assist you if you take a few minutes to go through the process.

Even a professional football team does not go out onto the field without going through some warm-ups and running a few rudimentary plays for a few minutes before the game. Don't you think the quarterback can throw the receiver a pass without doing so for a few minutes before the game? Of course. But going through some basics beforehand builds the flow, the confidence, and the momentum for a successful presentation. Do not shortcut these basics.

Get honest. Perhaps this is why presentations from politicians and bureaucrats are maddeningly boring. They can't be honest. They have to straddle the fence, to make all parties happy, and as a result of trying to please everyone they put everyone to sleep. Don't tiptoe around everything. Yes, there is a place for diplomacy, but generally being up front will be more appreciated than being namby-pamby.

Even people who disagree with you will appreciate the backbone that comes from unembarrassed integrity. A couple years ago I was lecturing in the agriculture college at a premier land-grant university and a very sincere grad student asked my opinion of what the USDA in general, and he in particular as a USDA official, could do to best move my agenda forward. I looked around at all the Ph.D.'s in the room and responded that I didn't think he really wanted to hear what I thought.

He disagreed. "No. Be brutally honest," he urged.

"Abolish the USDA," I said. You could hear a collective sucking in of breath in the room, but what followed was an incredibly healthy discussion of fundamental megapolitical issues, the role of government and personal empowerment. Although I did not convince any Ph.D.'s, my comments made a great impression on the students. In fact, one young man stayed later and said he would now rethink his goal to be a government employee and rather try to find some land in order to farm and lead by example.

Honesty moves people. Our world is full of mealy-mouthed scaredy-cats who wish and wash around every topic, saying all sorts of things and meaning nothing. An effective communicator gets down to the nitty-gritty and levels with people, looking them in the eye and sharing deep beliefs. This is what moves people.

Now that we've dealt with some of the communication basics, let me offer some tips for effective speaking and then effective writing. First, speaking:

Practice in front of a mirror. By far the best way to make sure your delivery is what it ought to be is to give it to yourself in front of a mirror, looking at yourself as you give it. If you can convince yourself, you can convince anybody. During my many years of forensics competition, I used this technique and it is incredible. It may sound hokey or elementary, but don't knock it until you've tried it. Looking at yourself in the mirror will help you with audience eye contact, gesturing, posture and expression. You can read tons of books on the subject, but nothing will improve your delivery faster than giving it to yourself in front of a mirror.

Listen to yourself on tape. Every "ya' know" and every "uh" will come out loud and clear. You won't need a whole lot of critiquing once you hear yourself. People have an innate sense of what makes an interesting presentation. Once you hear it, you will know if it is right or not. This includes enunciation problems, timing, grammar and volume. If you cannot clearly understand yourself, nobody else can.

Practice on your family. Your spouse and children will love to hear a presentation. We all like to be entertained, to enjoy a performance. Take their suggestions willingly and be quick to make changes. If you can make your family happy, you can make anyone happy. The people closest to you are generally your most picky critics. When you are finished, ask them to tell you in a nutshell what you said. If they missed your main point, you're probably not being clear. If they got it, probably everybody else will.

Tell stories. All audiences love stories. Illustrate your point with experiences or parables to give poignancy to your message. People love to listen to stories, especially ones with a little bit of humor. Permeate your presentation with stories and folks will stay on the edge of their seats. *Bartlett's Familiar Quotations is* a wonderful resource you can put on your next gift list. It's a compendium of famous lines, all arranged to help you find the appropriate one for your topic.

Now let's look at some hints for writing:

Eliminate these words from your writing vocabulary: *there, so, very*, and *it is* or *it was* at the beginning of a sentence.

Use active voice instead of passive voice. "The ball was thrown to John by Joe" should be "Joe threw the ball to John." Keep the verbs active so your writing has punch.

Rotate your sentence structure. Start with a clause and then put in the subject and verb. Don't just use the same simple structure over and over again. By diversifying your construction, you keep things interesting.

Look for prepositions and see how many you can eliminate. Generally, try to reduce the number of words. Talking and writing long is always easier than talking and writing short. "It's so very easy to have said too many words about the subject in the room above the cafeteria" could be condensed to "Room 223 encourages speakers' verboseness." Perhaps this is an exaggerated example, but you get the picture.

Submit your writing to a friend or relative who has a knack for the printed word. Often people who are not great writers are quite gifted editors. Many people do much better correcting someone else's writing than actually composing material from scratch. Perhaps one could be considered more an artist while the other is more a technician. The world needs both.

Look at the suggestions and ask for an explanation as to why your editor friend made the corrections. What you want to do is develop an eye for things like awkwardness and wordiness — some of these things are somewhat subjective, something you just feel when you look at your writing. All of us need people looking over our shoulders. We know what we want to say, but it doesn't always come across clearly. A second set of eyes looking at your manuscript is more beneficial than just about anything.

Now just some tips for specific types of writing you may be doing.

CORRESPONDENCE. Always reference the item to which you are responding in order to get everyone on the same wavelength right away. A letter is a paper trail, and the trail needs to be linked in order to have merit. Say: "I am responding to your *whatever dated such and such.* This provides a reference point for future use. Don't feel like you need to fill the page. Make it short and sweet. People want information, not pretty prose. As my high school journalism teacher always said: "A banana is a banana is a banana, not a long yellow fruit." You might write letters to people in the sustainable agriculture world to let them know you just read something they wrote or heard about what a good job they were doing. You might write letters of information to local politicians about something they discussed at a meeting and clearly did not know about your perspective.

LETTER TO THE EDITOR. This is a wonderful venue — use it every chance you can. Although large metropolitan papers have a competitive "getting in" situation, that is not the case in smaller newspapers. The highest reception rate is for letters responding to editorials; next highest is letters responding to specific news items, especially of local interest. Don't waste your time writing about world issues and generic national topics. Use a local current event as a springboard to your point and you'll have a high chance of catching the editor's eye. Make these terse and short — never longer than 150 words. The shorter, the better chance of getting in and the less chance of being edited or of losing the audience. Make a numbered list of your points — this reduces the likelihood of your letter being edited as well. If you get into the habit of writing once a month, be sure to write some that congratulate the editor on a position — this is equivalent to a bribe and will stand you in good stead down the line. Don't just write at election time. You can certainly write these about agricultural issues, using the opportunity to give yourself some free publicity.

BROCHURES. Leave lots of white space. Gray is too laborious to read. Be sure all the information is in there — what you're promoting, why they should patronize you, where you are, who you are. Always keep in mind the five W's and an H — who, what, where, when, why, and how. These are the questions you need to answer. Since these are introductory, they need to be more eye catching than informative. You're trying to reach someone who will only give you 30 seconds. Remember that. One of the most positive marketing tools you can use is the farm brochure. You can generate several different types for different audiences. Perhaps you want one focused on just one of your products, as an info-sheet. Go to the supermarket meat counter and notice some of the point-of-sale bulletins located there. The handouts are cheap to produce but highly effective in educating people, bringing them along in their thinking and thereby deepening their commitment to your product.

NEWSLETTERS. When communicating with patrons, give them lots of honest information. These folks are your core. These are your fans — they will stay with you. Without overdoing it, give these folks all the information you think they need to stay knowledgeable about your problems and victories. This is what will keep them up to speed with your problems, your plans and your products. Level with them and they will respond in kind. Depending on your farming enterprise, these can be generated as seldom as once a year or as often as once a week. Often CSA operations generate one a week to keep members aware of what is available. You can hardly communicate too much, especially with your core patrons.

Communication is crucial to every facet of life. Farmers need these skills as much as anyone and need to cultivate these principles to be effective. Don't get bogged down in the means until you have a clear message. Feed your passion until your cup spills over. At that point, your message will become magnetic, and that will create movement.

Pricing

Few things are as ticklish as setting a price. I am not a pure amoral capitalist whose simplistic answer is: "Get as much as you can." Life consists of more than just making money. We should not even attempt to become wealthy. By the same token, neither should we sell ourselves short. Generally I think farmers price themselves far too low instead of too high.

I like to start by establishing a baseline return per hour. Dad's famous phrase, "You might as well do nothing for nothing as something for nothing" is certainly true. If I'm not going to be decently compensated, I may as well take a no-risk job in town that I can leave any time I want. I don't have to think about it until 8 a.m. and can walk away from it at 5 p.m. Someone else takes all the risk.

Many people live that way and have no aspirations for anything better. But if you do have higher aspirations, then working for nothing is not a satisfactory option. If I'm going to take the responsibility for a piece of ground and a small business, then I do not think it arrogant to seek professional-status compensation.

We all have 24 hours a day. Time is the great limiting factor for us; it is also a great equalizer. All anyone has to market to society is a product, idea or service. Obviously, if we can market all three we get a cumulative effect for our time.

Photo 37-1. *Rachel works on an invoice in the on-farm sales building. Being a price maker instead of a price taker is one of the concept pillars in entrepreneurial farming.*

But if we start with the question: "What are you willing to work for?" we at least have a benchmark. How much per hour would you charge me to hire you? Any self-respecting person is not going to say: "Minimum wage."

Let's not get bogged down in a moral discussion about how much a person is worth. I know people who charge $200 per hour for their time. I had a surgeon once sew up my finger in the hospital emergency room and charged $600 for 30 minutes' work. Let's not, for our discussion, deal with that scenario. Let's look at compensation realistically. If you could get $10 per hour, would that be satisfactory, at least for starters? How about $15? Would you need $20? You see, what you want is not the issue. The issue is to choose something as a starting point.

If you watch your costs going into the product, then the difference between your input and gross sale price is what you have left over to pay yourself. I know some folks will take issue with me here because I'm not putting in every possible item, like the cost of the hoe, the pruner and the wheelbarrow. But most of us would have those things anyway. We're just trying to get a handle on a concept here, not debate all the finer points.

Adjust your price so that it gives you the return per hour that you've suggested is your threshold of satisfaction.

For example, we've generally established $20-$25 per hour for our return. That seems quite satisfactory to me. We rent a farm about 12 miles away and run about 80 stocker calves over there. We spend two hours every other day moving them to a new paddock and setting up or taking down fence. We buy them in early March and sell them in late October, which amounts to seven months' ownership. In that time, we have roughly 100 trips over there at 2 hours per trip, for a total of 200 hours. After we take off trucking costs, supplies, minerals, mileage, and rent, if we end up with $8,000 net, we figure we have about a $40 per hour return to labor. I know I should account for what we could have gotten on the money if we had put it in the bank at interest, but again, we're trying to get general ideas here. Obviously, that return meets our minimum return to labor criteria.

If I spent 5 days sitting at the stockyard purchasing those 80 calves, those additional 30-40 hours would significantly cut into my return per hour. That is why I use an order buyer. My time is far too valuable to be squandered sitting in the sale barn. He's there anyway, so he doesn't attach a significant cost to purchasing my calves. Since the cost is in being there, it's better for me to use his trip and for me to stay home and use the telephone.

Let's say you're selling strawberries. How long does it take to pick a quart? If it takes 10 minutes and you can pick 6 quarts per hour, you would need to clear $3 per quart to get $18 per hour return to labor.

This, just like accounting, can help you decide what enterprises to pursue and which ones to stop. It also shows the foolishness of the industrial model. Why do more farmers with factory confinement houses also work at a job in town or send their spouse off the farm for additional income? If those things are really moneymakers, why doesn't anyone stay home and build more?

My friend Jeff Ishee recently had dinner with a confinement poultry farming couple at a banquet. He reported: "We were talking about women in agriculture. I raised the point that one of every five new farmers in Virginia is a woman. Then I asked why that was happening. The lady looked at me and said, 'Well, that's no surprise. My husband got a job at the factory in town, so on this year's tax return I was listed as the principle operator of the farm. There it is ... I'm your new farmer.'"

I know many farmers who have bought into the industrial paradigm and not a single one of them is really making money. They are making payments. They might get minimum wage. But almost none is really earning a white-collar equivalent.

Keeping a record of time spent on task is crucial to pricing. For example, we know that it takes 3 person-minutes per day to move and service a chicken pen. If we dress 80 birds out of that pen, we have put 105 person minutes in that pen during the five weeks they are on pasture. Add on another 5 for the brooder phase and you have a total of 110 minutes per 80 birds, or 1.375 minutes per bird. With our processing setup, a 5-person crew can dress 120 birds per hour, which translates to 24 birds per person-hour or roughly 2.5 person-minutes per bird. If you add the two together you get in the neighborhood of 4 person-minutes per bird. Our gross margin per bird runs like this:

chick	.60
feed	1.50
utilities	.10
supplies	.05
Total	2.25

If each bird averages 4 pounds dressed weight, our break-even is 57 cents a pound. If production and processing take 4 person-minutes per bird, we can do 15 per person-hour. If we want to make $20 per hour, we need to get an additional $1.33 per bird, which divided by 4 pounds would mean we need to add 33 cents minimum to the break-even rate. That puts us at 90 cents per pound.

Whenever you do calculations like this, add in a huge cushion because myriad unforeseen things will happen. A snake will get in and kill 20 chicks. You'll run over a couple with the dolly. A rainstorm will come and drown 30 of them. Of course, these catastrophes won't happen every time, but they do happen and you need to give yourself a good cushion. So let's say we just go ahead and add a quarter a pound just to be plenty safe. Now we're starting to get an idea of what we need to charge.

Actually, our price is $1.50 per pound but that has come after charging the minimum for a long time. You have to start somewhere. I know some people say you should start high, but I think it's better to start a little low and get your foot in the door, build up some loyal customers, and then gradually move the price up. If you start real high at the beginning, you'll have a harder time getting folks to buy. Even for clean foods, people are quite price-conscious.

In the ideal world, you would get paid for every hour. But in the real world, you do not. The goal of any beginning farmer is to become fully employed. As your market builds for the things that do return a good wage, then you can begin paring down the things you've been doing with your time that do not return that amount. The trick is to gradually move away from non-paying hours and get all your hours to pay at that established rate.

With all this in mind, remember that everyone has salvage times. In the fresh fruit and vegetable business this is normal. One of the best vegetable marketers I know, Chip Planck of Wheatland Vegetable Farms in Virginia, says that when you get to the end of the day you just have to drop the price to whatever it takes to move the stuff. He is exactly right. Inventory adjustments are normal and if you have to drop the price down to fire sale rates in order to move something, that's okay. If you need to do that, however, try to use it to marketing advantage.

I've generally found that it is better to give away product than it is to drop the price way down. Holding the line on the price and offering your overage as an incentive to buy other things that are not in short supply keeps you from compromising your price.

For example, we overestimated the amount of ground beef people at the farmers' market would buy and came down toward the end of the season with way too much in the freezer. We ran a special deal. "Buy $20 worth of merchandise and receive a free pound of ground beef." This stimulated people to buy our regularly-priced products — in fact, more than they normally would — and offered the goodwill and public relations of getting something for nothing.

People love to get something free. If you drop the price to a salvage rate, be sure to explain why you're doing it. Otherwise people may think you have damaged or tainted stuff and the sale will hurt more than help. We ended up dropping the ground beef price way down and put up a sign that read: "Help us make room for premium steaks and roasts. We need freezer space."

This ad gave people the impression they were helping us. People love to feel helpful. Further, it made all the folks who were waiting for steaks realize they could speed up the turnaround to a better inventory if they went ahead and ate meat loaf for a couple of weeks. It also communicated the real problem, and you can never overcommunicate with customers.

Anytime we raise the price, we tell folks why. As a general rule, use general information to raise prices, but never lower them. A few years ago when grain prices spiked up it made headlines in the news. We used that opportunity to raise our chicken prices 20 cents a pound, which was a huge increase. We were afraid of the repercussions, but everyone was aware of the nationwide problem and understood our predicament. Of course, when prices dropped back down to pre-crisis levels, we never dropped the price.

If you will piggyback your increases onto generally known

problems, you will find people quite willing to go along. People hate to see prices go up for no apparent reason and without an explanation. If you'll level with folks, they will come along with you.

Be sure to acquaint your folks with volume savings. For example, our beef by the quarter, half and whole is priced about 30 percent higher than beef purchased that way from a butcher shop or supermarket. But if you buy it by the cut over the supermarket meat counter, our volume price is about 10 percent less than the top quality equivalent.

Therefore, when we discuss pricing, we do not compare ourselves to halves and quarters, but rather we compare our volume price to the by-the-cut price at the supermarket and show how folks can save 10 percent on their beef by buying ours. How you position your competition relative to your product is up to you. Define the comparison in terms that are honest, but in terms that make you shine. Folks understand volume discounting, so they are naturally inclined to go that route when we focus their attention on it.

Now that we've dealt with pricing details, let's look at the whole issue of pricing for a premium product, including food pricing in general.

"You get what you pay for." We use that phrase all the time. We use it for boom boxes, clothing and services. We use it for automobiles, diamonds and houses. We use it for appliances, teachers and computers. There's only one thing in our society that we exempt from the saying, and it is ***food***.

The industrial agriculture sector continually bombards us with the notion that all pork is pork. All chicken is chicken. All eggs are eggs. All strawberries are strawberries. To suggest that wide fluctuations exist is to question Mom, baseball and apple pie. It is downright un-American to suggest that one egg might be nutritionally superior to another.

We must proclaim it long and hard: "You get what you pay for! And that includes food." We all know, innately, that food quality varies widely. Yet the essence of industrialism, which is standardization, cannot allow for comparisons. That is why the FDA and USDA refused to allow organic milk processors to label their milk "BGH-Free." It casts disparagement on all the other milk that is *not* full of pumped-in hormones. [A recent lawsuit by Ben & Jerry's may change the labelling rules.]

Genetic engineering is enjoying the same protection. Food on the grocery store shelves that contains genetically engineered ingredients is not labeled. Chicken that is 10 percent fecal soup is labeled the same as clean-processed, range-raised chicken. Of course, the products are dramatically different but our society, conditioned to love standardization and cheap food, has a real problem accepting comparisons and higher prices.

Look what cheap food has gotten us. The societal costs to clean up the manure spills, to treat the sick and dead from food borne bacteria, to handle and inhale the mountains of manure from centralized confinement facilities are astronomical. If all farmers would do things right, the people employed in government and private agencies to advise them about animal sickness and pollution could go out and do something productive instead of parasitizing our economy.

A cheap food policy encourages corner-cutting. When the poultry industry feeds chicken guts to chickens because it represents a cheap source of protein, you can blame the selfishness of a cheap food policy. When we turn herbivores into carnivores by grinding up and feeding dead cows to cows, you can blame a cheap food policy.

Consumers and farmers share the blame equally. This is not the time for excuses. It is time, instead, to take an unapologetic stand for pricing food where it ought to be priced. And that means that cheap food is priced about right — it's worthless. Not only is it devoid of nutrition, but it actually makes people sick. The Government Accounting Office, using figures supplied by the Centers for Disease Control, estimates that up to 5,000 Americans die every year from food-borne illness. And remember that according to the General Accounting Office, the culprits are large-scale processing facilities, centralization in the production sector, transportation and long distances between producer and consumer, and virulent mutated or resistant strains of pathogens due to routine medication or other toxic materials.

A large sector of the population is waking up, beginning to take responsibility for correcting this policy. These folks are the ones who take the time to educate themselves and they will carry your farm to success.

When someone complains about our price, I like to use a phrase developed by Jeff Ishee, author of *Dynamic Farmers' Marketing:* "That's what it cost me to grow it." It is a disarming response. First, the customer has no way of knowing how much it cost you to grow it, so to argue will expose his ignorance. Jimmy Dean said: "I'd rather explain price than apologize for quality."

Second, it is a gentle way of saying: "I'm worth something." My position is this: "If you want to smell from my farm what you smell from most farms, then go eat fecal soup. But if you want me to stick around and be in business 10 years from now producing environmentally enhancing, nutritious food, then you'll think this is a bargain."

We farmers must break through the notion that we're supposed to be peasants out here working for nothing so city folks can drive around in BMWs. Nonsense. Any city person who doesn't think I deserve a white-collar salary doesn't deserve my special food. Let them eat *E. coli.* That sounds harsh, but I simply have little patience for someone who disrespects me that much.

What is really the kicker is when a lady stands there complaining about $2 a dozen for eggs while she's sipping a Coca-Cola. The biggest food companies in the world do not produce food. Remember that. Taco Bell, McDonald's, Burger King, Pepsi and Kentucky Fried Chicken are not in the food business. They are in the entertainment and recreation business. Never forget that. They are not selling nutrition. Just one of my eggs has more nutrition in it than a carload of soft drinks. And yet this lady has the audacity to tell me my eggs are too expensive. People have money for whatever they think is important. I do not believe we should pass laws about how people spend their money. But neither should you let a consumer light a cigarette and tell you prices are too high. Wrong answer. The right answer is that they have money for whatever they deem important. It's a matter of priorities. Do not apologize for pricing your products in such a way that all the costs are covered and you have a return that will make you want to stay in business the rest of your life.

Here are two sample interchanges I've used to kindly educate people about this issue.

CUSTOMER:	Ground beef for $3.00 per pound? That seems high.
ME:	Well, one pound goes as far as two from the store.
CUSTOMER:	How can that be?
ME:	Ours is leaner because the cows eat green pasture grass instead of corn, dead cows and chicken manure. We don't put hormones in their ears either.
CUSTOMER:	No, really? Cows eat that stuff?
ME:	Yes, ma'am, state-of-the-art, right out of the extension service. They've got to do something with all the waste from factory farms.
CUSTOMER:	My, my, I think I'll take two pounds.

****

CUSTOMER:	Why is your chicken so high?
ME:	Have you ever been in a chicken factory?
CUSTOMER:	Well, no, but I've seen them.
ME:	Have you ever smelled one?
CUSTOMER:	Oh, yes, they're awful.
ME:	Who pays for that smell?
CUSTOMER:	I guess we all do. Sure stinks up things, doesn't it? You know, I read in the paper last week about some neighbors suing one of those farmers for contaminating their wells.
ME:	Who do you think pays for that?
CUSTOMER:	We all do. I guess the government will have to

go in and try to clean it up.

ME:	You're right. By the way, have you ever seen *60 Minutes*?
CUSTOMER:	Oh yes, that's one of my favorites. They had a piece on their recently about people getting deathly sick from eating dirty chicken. It was awful.
ME:	Yeah, I'm sure glad the chicken companies paid those doctor bills.
CUSTOMER:	Why, you know they did no such thing. Those companies didn't pay anything to those poor people.
ME:	Begging your pardon, ma'am, but our chicken won't hurt anyone; it won't destroy anyone's backyard barbecue this afternoon, and it won't pollute anybody's well. All the costs are in there. Nobody else is going to pick up the tab for what you don't pay.
CUSTOMER:	You are right. I never thought of it that way before. Give me two.

Use your pricing to heighten awareness and create more loyalty for your product. People generally are woefully ignorant. You would think everyone knew they feed chicken guts to chickens, but they don't. Everyone is provincial in a way. We all need to be reminded about things that are not in our daily routine, our little world.

If your quality matches your price, you will find people affirming your position. You do not want customers who buy from you because you're cheaper than anyone else. You want folks patronizing you because you're better than anyone else or because they can't get what you're selling anywhere else.

I would hate to go through life producing and selling something in which my marketing angle was that I was cheaper than all the other options. I would find that ungratifying. To garner market share just because you have a cheaper product, it seems to me, is an unfulfilling mission to live for. But being able to stand tall and proclaim, “This is the best tomato in the world” — now that is a noble cause. That is worth living for. Nothing else is noble enough to consume a lifetime.

If your product merits its price, you will find people gladly rising to your standard. Wouldn’t it be exciting to develop a food system based on excellence instead of mediocrity? Do not try to gouge, but hold your head high and set your price at a self-respecting level. Nevertheless, stay flexible. If nobody buys, you’re probably too high. Be reasonable and let pricing be a way to show your integrity and earn respect.

Value Added

Farmers should try to capture as much of the retail dollar as possible. From the farm to the store, farm products get trucked, processed, advertised, inspected, labeled, warehoused, packaged and carted, then finally scanned by computers at the supermarket checkout counter. Each of those procedures adds incrementally to the final retail price.

Just three decades ago farmers received 35 cents of the retail dollar. Today they receive less than 9 cents and the line is trending downward. Although different studies use different numbers for starting and ending, all agree on the general size of the difference and the general trend. This reflects several changes in the farm-food chain.

First, farmers are doing less of the processing and marketing. In fact, on the most state-of-the-art farms, the farmer does not even own the livestock he grows. Some huge corporation owns the animals and contracts with the farmer to grow them. As the farmer gives up more and more of his control, his share of the food value decreases.

Farmers used to cure pork and sell it to local restaurants. Chicken producers used to pedal their eggs through town. Dairies used to deliver fresh milk right to your doorstep. By the way, some dairies are beginning to do that again and they are being quite successful.

The actual raw material producers have less and less stake in the final product. It is only fair that the middlemen all the way up the chain get paid for their contribution to the product. Farmers used to store most of their grain on the farm and sell it as they needed cash. Now they just take it all to the elevator and take what they can get.

Another reason for this drastic decline in farmer share of the retail dollar is the working woman. Three decades ago most women were still at home cooking food from scratch. But with women joining the workforce, more meals are now prepared outside the home. Ten percent of all food is consumed in automobiles.

Instead of buying Idaho baking potatoes for 9 cents a pound, taking them home and cooking them for the family, the consumer buys microwave-able French fries for 91 cents a pound. I am amazed every time I go into a supermarket and see the prices on processed food. There is frozen pizza, precooked chicken, precooked bacon at $20 per pound. The list is endless. Roughly 50 percent of all meals are now prepared outside the home. This includes frozen TV dinners as well as restaurant meals. Most banquet meals now come frozen in boxes. The cooks just pop them into huge ovens and heat them up. These ovens have multiple narrow sliding drawers for stacking the meals.

Nearly 70 percent of turkeys are further processed into turkey ham, turkey franks and turkey bologna. Canned soups, vegetables and fruits occupy huge shares of supermarket space. The price of all these highly processed foods compared to the raw thing is nothing short of amazing.

The figures factory farmers give are equally amazing. They talk about getting less than a nickel per chicken. The margin on these industrial models is so small it's a wonder any of these guys makes a dime. Some turkey growers now say that the workers on the catching crew make more money than the farmers do. But when you voluntarily give up everything except the mortgage, the light bill, and hauling manure, you are guaranteed a miniscule piece of the pie.

Photo 38-1. *Pastured poultry spin in a farm-sized automatic chicken picker as one step in the chain that adds value by making an oven-ready product. Processing and marketing are where the real dollars are, not in wholesale pricing.*

Why more farmers can't see this is beyond me. They're still lining up to bite on this hook dangling enticingly from the agribusiness pole.

I'm not satisfied with 9 cents on the dollar, and neither should you be. The only way to play the wholesale market game is with volume, and very few people can afford the capital to play that game. For most of us, a more realistic approach is to make a decent net return on each animal, each carrot or each piece of firewood. That means getting as close as possible to the retail price.

When we take a load of logs to the sawmill, a fellow comes out of that scalehouse who will tell us how many board feet we have, how good the logs are and what the price is. Somehow that doesn't put me in a frame of mind to negotiate from a position of strength. We farmers have become price takers instead of price makers, and it's time to turn that around.

But it won't come by demanding a fair price, or driving our tractors to Washington, or lobbying for parity. It will come only as you decide to do something more with what you produce than lay it at the mercy of food processors. They will take your carrots or lamb chops and make them unfit for human consumption by the time they amalgamate, extrude, procrastinate and adulterate them.

The whole idea of value adding is to do something besides sell it like everyone else. Sometimes value adding can be as simple as growing a special variety, like blue corn instead of yellow. The specialty item commands a much higher price and is usually not that much more difficult to grow. If you can double the price, even though

Photo 38-2. *Rachel takes a package of "Pigaerator Pork" out of the sales building refrigerator. The difference between wholesale and retail averages about 300 percent on pork.*

the production is 20 percent per acre less, you've increased your income on that square footage by 80 percent.

Remember we are in the designer age. Specialty anything attracts attention. Of course, as soon as you go to specialty products, you will begin looking at unconventional marketing. Large buyers are looking for standard fare. Meat packers want black cattle because they fit the box. Tomato packers want certain varieties. Don't try to sell blue corn down at the Archer Daniels Midland grain elevator.

You can always find a market for a specialty product. Who would have thought a guy could paint eyes on a rock and sell it . . . by the millions? There's always room for a new idea. If you're going to take a bunch of product somewhere, it may as well be $3 per pound material instead of $1 per pound.

Perhaps you are not interested so much in specialty as you are in some further processing. For example, imagine the value of an acre of corn sold as cornbread muffins. You could make as much off half an acre of corn sold that way as a farm with 1,000 acres would make selling it to Pillsbury. This allows you to put more of your energy on non-capital business elements. Rather than having to own the land and equipment to produce hundreds of acres of corn, you simply use your labor to make cornbread muffins. You need not borrow your labor from a bank. You need not mortgage your ingenuity.

Our son Daniel taps half a dozen sugar maple trees in the yard and on a neighbor's land and boils the sap down to syrup. In our area maple syrup sells for about $45 per gallon. He makes a couple of gallons. Rather than sell the syrup, he bakes maple donuts and sells them at the farmers' market for $4 a dozen. This translates into almost $200 a gallon for the syrup.

What he's doing is taking a small production and adding his labor and creativity to it in order to make it worth more. This is what every giant food processor does and there is no reason why more farmers can't do the same.

Value adding increases the number of people interested in

the product, which necessarily spreads your market base. When you take a steer to the stockyard, very few people are interested in bidding on that critter. But if you offer T-bone steaks, suddenly the world is ready to buy.

It only makes sense that the more buyers you have, the more potential you have to get a fair price. Selling wholesale limits you to buyers who have the wherewithal to take your raw material and turn it into something consumers will want. You can do it just as well and capture all that labor cost.

Have you ever noticed how wholesale prices fluctuate drastically but retail prices stay relatively steady? Beef prices at the stockyard may fluctuate 100 percent within one year, but they vary scarcely 5 percent in the supermarket.

If the boom and bust stories farmers had to tell were all written down the Library of Congress couldn't hold all the pages. We're all familiar with a shortage year causing high prices. Then the farmers all plant that crop to capture the high prices, which causes a glut and knocks the prices lower than they ever were. Selling retail insulates you from these wild fluctuations.

What do farmers sit around and complain about? Every rural diner in America has a table of these 60-year-old codgers sitting there grumbling about weather, disease, pestilence and price. Those are the four big variables of farming. Have you ever heard the farmer likened to a gambler because of these variables? Of course. "Farmers are the biggest gamblers in the world," the statement goes.

Those variables, however, are really only accurate on wholesale production. Value adding reduces the impact of those variables. The more farm income is subject to those variables, the more precarious the balance sheet.

If a farm derives a significant portion of its income dollars from marketing and processing — value adding — rather than wholesale production, it becomes much more stable as a business. Direct marketing and processing are not subject to weather, disease, pestilence and price. If all the farm dollars derive from wholesale corn production, a drought or pest infestation can wipe out the profit.

But if half the dollars come from marketing and processing, at least those dollars are protected from the drought or infestation. A woodcutter knows that good chainsaws are 80 percent sharpness of chain and 20 percent machine and operator. By the same token, profitable entrepreneurial farming is 80 percent good marketing and 20 percent farming.

Let's look at it another way. A corn farmer needs 1,000 acres to make a living. A drought strikes. He's borrowed money at the bank to put in the crop. He watches the skies each day and then watches his crop wilt in the field. It's too massive to do anything about, so he just shrugs his shoulders and takes his beating.

His neighbor has 5 acres of corn that his family turns into cornbread muffins in their inspected kitchen. The drought comes. The man and his wife and their three children rig up garden hoses to the well. The children take turns putting a gallon of water on each plant because they don't know what the well can take. They can't get to it all from the well so they go down to the rental agency and rent a pump to put in the little stream behind the house. They finish the 5 acres in a week and the corn grows straight and tall.

They have a fine crop and they begin making muffins. Because most of the income was from an enterprise that was not subject to the vagaries of weather, disease, pestilence and price, salvaging the amount that was subject to those variables was do-able.

The same scenario could be offered for livestock. Compare trying to salvage the pigs out of a confinement hog factory in Iowa during those tragic floods a few years back to salvaging 20 or 30 from a field. The small group could probably be herded up onto a hilltop to wait out the flood. They would probably just walk to higher ground by themselves. But most farmers were trying to load pigs into little Coast Guard skiffs. It was impossible. Nearly all the pigs were lost. What a tragedy.

This is one reason why I encourage people who want to go out and buy a farm to put their money in an inspected kitchen instead. Start making baked goods, breads, vegetable and meat pot pies. That $10,000 investment will pay back much quicker than the same money put in land. Once you have a loyal client base, you can begin integrating into growing more of your own ingredients.

Then you can buy your land, move the kitchen, and maintain your clientele. It's all a sensible, smooth transition. People put so much romance in owning that acreage, though, it's awfully hard to realize that a farm is only a subset of the food system. A farmer is not sacred. If all he does is produce raw wholesale material, he is only the first link in a long chain before that food gets to the consumer.

The farmer is inherently no more important than any other link in that chain. I know this will get me into trouble with many folks, but I believe this unnecessary veneration has skewed farmers' minds. They think society owes them something just because they are farmers. We farmers tend to think we have more risk than other businesses.

Hogwash! Farming is no more risky than selling pots and pans or fixing shoes. Nature operates by a set of principles just like any business. The business manager who must assess market directions, determine prices, maintain inventory and advertise needs just as much savvy and is subject to the same temptations and mistakes as any farmer.

The reason we farmers make ourselves holy is to create the martyr's complex, to get sympathy and accolades from people. We say we're completely at the mercy of nature, and we are, but nature operates based on known principles. We mess things up by putting orchards in frost pockets. We rip up the perennial prairie polycultures, exposing the black, light loam to the sun, the wind and the torrents. And then we blame nature for our catastrophes?

Please, please. If we would quit shaking our fist in God's face perhaps we would receive more mercy. If we would adopt more business-like models, perhaps our farm businesses would be more prosperous and less precarious.

Loading our farm income dollars into marketing and processing stabilizes our income, spreads our risk and helps to protect us from weather, disease, pestilence and price. We need not age into

Photo 38-3. *Rot-resistant pole timber, like these locust trees, are far more valuable as building poles than as cordwood or pulpwood.*

Photo 38-4. *Some of the greatest value you can add to your own farm production is utilizing your own crude resources to replace materials on which other businesses — including most farmers — spend millions. This 5,000 square foot shed awning only cost $2,500 in out-of-pocket expenses and yet it is as functional as any fancy "boughten" variety.*

bitter, grumpy old farmers. We can actually age with a great appreciation for the limitations of nature and the simple elegance of applied business sense.

Value adding offers unparalleled opportunities for turning small acreages into big bucks. Whatever product you think about growing, think about capturing retail dollars. If you raise sheep, think about sheepskins, lamb chops and yarn. If you grow carrots, think about baby carrots, sliced carrots and vegetable pot pies. If you have an especially pretty setting, think about marketing the scenery through farm and ranch vacations, fee hunting, or bed and breakfast.

In short, value adding is the ticket to a white-collar salary for a small or beginning farmer. I have not yet met a farmer who could not value add something. Certainly some areas are more conducive to further value than others are, but if every farmer would move toward some of these principles it would fundamentally change agriculture in America.

All over the country farmers attend conferences to hear speakers lambaste this government agency and that corporation. They sit and listen for three days to speakers complain about things not a one of them can do anything about. A man with 40 cows can't change the buying policies of Excel. One poultry grower can't take on Don Tyson. A 30-acre green bean farm can't take on the Jolly Green Giant. Certainly no farmer can shut down Monsanto or Ciba Geigy.

But we as entrepreneurial farmers can take the reins of our own destiny and begin a steady, systematic journey toward opting out. We just refuse to participate in their game. We can either spend our energy complaining about things or making creative changes. Too many of us have tried the first path. I suggest we try the second one.

You may not change your neighbor. You may not change the feed store. But you can change you. And that's what matters. Let's dedicate ourselves to getting as much of the retail dollar as we can get and let the big boys go play their games. We're on the road to success.

Summary

In Summary

For several hundred pages now I have shared my soul with you. I hope my attitude in presenting this material has been one of encouragement and helpfulness rather than arrogance and judgment. I do not have all the answers — I learn something new every day. Even though I do more speaking and writing now than I used to, I still thoroughly enjoy moving chicken pens, cutting firewood and moving cattle. I have no intention of seeing my heavily callused hands soften through disuse.

As entrepreneurs, our family created and now enjoys a thriving business that appears stable for decades to come. While duplication is the highest form of compliment, you must remember that an entrepreneur creates something brand new, something that has never existed before. The way the business looked when it started is generally not the way it looks several years down the road. Your farming enterprise will be unique; it will be an outgrowth of who you are and what you believe in.

Several folks around the country have already quite successfully duplicated many principles I've espoused in this book. In fact, some folks actually use language from newsletters we've written in their marketing efforts. That's fine. But what is fascinating to me is that when I visit with these folks, no one would confuse them with

Polyface. Each one has enough different pieces that it shines alone in the community.

Common principle is foundational, yes, but it is still only common principle. The template allows enough individualized fleshing-out that every current farmer or wannabe can create something brand new, something distinctive and rewarding.

The broad concepts I've outlined here work in every part of the world, even transcending time. If these principles had been adopted broadly a century ago, our culture would look much different now. We would be a different people, but we would still be as individualized as we are now. I'm always amazed that people are still making up new titles for country music songs. You would have thought all the possible ways to say you broke up with your girlfriend or boyfriend, lost your job or your truck, or looked for fulfillment, would have been said. But apparently enough word combinations exist to keep Nashville going for a long time.

I have no illusions that this book will transform our culture, as wonderful as that would be. It will be enough if it touches you and makes you a more enthusiastic, creative, caring, honest person.

Everyone has different experiences to share. Everyone comes at problems from a different perspective. I have simply articulated my experiences and my perspective. You will not duplicate them. You cannot. You may have similar — strikingly similar — ones, and if I didn't think you could I would never try to encourage you to "go thou and do likewise."

I do believe strongly that our world is full of people who could enjoy higher expectations, higher callings for themselves and their families. You may be one of them. If you are, my prayer is that you will be committed to truth, that you will be dedicated to people, and that you will be reverent toward God's creation. If enough of us will accept this challenge, we can make a difference. *Let's do it.*

Appendices

Newsletter

SALAD BAR BEEF
PASTURED POULTRY
PASTURED RABBIT

PIGAERATOR PORK
FORESTRY PRODUCTS
CONFERENCE SPEAKING

Polyface Inc.

"Farm of Many Faces"

LUCILLE SALATIN
(540) 885-1166

JOEL F., TERESA W.,
DANIEL & RACHEL SALATIN
(540) 885-3590

RT. 1, BOX 281, SWOOPE, VA 24479

Spring 1997

Dear friends,

Have you noticed the new supermarket names? **KROGER FOOD AND DRUG. GIANT FOOD AND DRUG.** As we've adulterated and refined all the nutrition out of our food, we've become pharmaceutically-dependent to the point that even our grocery stores partner with the drug companies in order to prop up patrons long enough to get them through the check-out lines.

Perhaps no writer articulated or prophesied our current situation better than Sir Albert Howard, "the father of composting," who penned these words in his 1940 classic, *AN AGRICULTURAL TESTAMENT*: "Artificial manures lead inevitably to artificial nutrition, artificial food, artificial animals, and finally to artificial men and women."

Think about the last "diet and nutrition" article you read. What are the basic tenets of any honest nutrition program? Here is the generic agenda:

1. Eat fresh
2. Eat local
3. Eat unprocessed

Certainly other elements come to mind, but these are the basic points you'll hear over and over. As you know, we offer this type of food.

Do you realize that the largest "food" companies in the world do not sell food based on the above goals? Coca Cola, Pepsi Cola, McDonald's, Kentucky Fried Chicken, Taco Bell and the others sell entertainment, convenience and recreation, not primarily food. Our society confuses the two.

I've had folks complain about paying $1.50 per dozen for eggs while sipping on a 75-cent can of soda. One egg contains more nutrition than a can of soda. Here at Polyface, we are in the nutrition business. Of course, we want you to have a good time when you come, but we are primarily in the clean food, clean farm business. You just can't get it any more fresh, local or unprocessed.

Economics and the environment go hand in hand. Compare the price of packaged, frozen ready-to-cook French fries at 91 cents a pound with premium grade Idaho baking potatoes at 9 cents a pound. The other day I found pre-cooked bacon in a box at $4.99 per 4 ozs.—that's $20 a pound for bacon! And somebody is buying that stuff, or the supermarket wouldn't have it occupying shelf space.

When you accept some of the responsibility for handling, storing and preparing your food, you can afford top quality. Don't let anybody kid you into thinking amalgamated, extruded, irradiated, genetically engineered, adulterated fecal particulate pseudo-food from Archer Daniels Midland is economically or nutritionally viable. Researchers watch mice get sick on a breakfast cereal diet, but watch them get healthy when fed the raw ingredients listed on the label.

A Government Accounting Office study released last year identified centralized food production and processing systems as the primary culprit in the burgeoning cases of food-borne bacterial illnesses in the U.S., which in 1995 soared to more than 5,000 deaths. Centers for Disease Control now estimate that half of all cases of diarrhea are caused by food-borne bacteria. When a truck hauling chicken wings today hauls fruit juice tomorrow, and bags of dehydrated slaughterhouse wastes the next day, you're asking for trouble.

When you couple the direct health risks of conventional fare to the environmental degradation caused by corner-cutting industrial farms that receive special concessions by cozying up to politicians, you realize how important it is to opt out of the mega-food system. The poultry industry is in a dither over new standards that will require them to wash off all visible manure from carcasses **before** the birds go into chill tanks. Now we know why TIME magazine in 1994 reported that chill tanks had 12 inches of fecal sludge in the bottom and up to 10 percent of the weight on supermarket birds was fecal soup insoaked during the chilling process.

All of this and more caused Charlie Walters, editor of *ACRES USA* magazine, to exclaim recently: "At current government toxicity levels for food, if we lived in a cannibalistic society, none of us would be fit to eat." The food-farm link deserves some attention in our culture. As our culture rushes pell-mell into a faster pace demanding more recreation, entertainment and convenience, you are enjoying the emotional, economic and environmental benefits that come with heading a different direction. We at Polyface are proud of you, and honored to have you along with us in this journey toward truth. We look forward to celebrating this year together as we chart a different course. In the words of Robert Frost:

"Two roads diverged in a wood, and I-
I took the one less traveled by,
And that has made all the difference."

PASTURED BROILERS Last year's cold and wet season was tough on chickens but great for cows. That's one of the advantages to being diversified. We realize that the weights were low, and apologize for that. Weather is our biggest variable; not every day is 70 degrees with light breezes and azure skies. Hopefully this year will be better. We are doing some research trials that we hope will help as well.

Each year the industry moves farther away from reason and nature in its production models and genetic selection. This makes it more difficult for us to raise the chickens and we are trying some high-tech biologicals like *Immuno-boost* and mineral and enzyme products to augment our natural feed ration. We'll look forward to your feedback as the season progresses. Price is the same as last year.

One of our patrons last year subjected our birds to a battery of radionics tests, measuring for toxicity and nutrition, especially B vitamins and minerals. Not only were toxins absent, but nutrients were much higher than in supermarket birds. The conclusion: "Eating one of your birds is like taking a vitamin pill." How about that?

This year we ask that you circle a processing day on the order blank even though it may need to be changed later. This will keep us from having to make so many phone calls. We'll just send a postcard reminder and if the date you picked does not suit, you can always change it up to 24 hours in advance. We had some trouble with the Lexington area especially last year receiving post cards. Hopefully that will not happen this year. As usual, we're on a first come, first served policy so when the first batch fills up, we'll bump you to the next batch and so on. Remember that the first batch to fill up is always the last one of the year because everyone balloons their numbers for the winter.

PASTURED STEWING HENS Many of you purchased or received an introductory one free from us last year, and the feedback is quite positive. These are laying hens that have finished two years' active production. While they are not as tender as the broilers, the taste is out of this world. If cooked long and slow, they are tender. We cook several at a time in a large roaster, pick off the meat, dice it up and freeze it. This precooked chicken is then ready for casseroles on a moment's notice. It takes less freezer space, can be done on "off days," and is handy. The broth is incomparably superb. We have a freezer full from late last fall if you'd like some now. It will be a couple of months before we have fresh ones. Prices same as last year.

SALAD BAR BEEF A new study completed in Australia concluded that this kind of beef contains 300 percent more B vitamins than grain-fed beef. Of course, the fat is different both in volume and content, but the taste is superior to feedlot beef. As more sophisticated measuring tools develop, we simply grin at the "new findings" which prove that what we've been doing for decades is best, The confirmations are fun.

By far and away the easiest and most economical way to purchase this is by the split half, side or whole. We do, however, have it by the cut, processed under federal inspection which adds about 20 percent to the cost, but allows you to get it by the piece. It is here at the farm now if you need some to tide you over. A split half, which is half of hind and half of a front quarter, is not that much volume on a smaller beef. The order blank gives you the option of large and small animals. One on the small side would yield enough take-home packaged beef to only fill two bushel baskets—that's not a whole lot. If it's still too much for you, invite a neighbor or friend to go in with you. Because we mimic natural herbivore populations with our grazing management, you can participate in regaining perennial prairie polycultures by eating this environmentally-enhancing beef. Prices same as last year,

PIGAERATOR PORK We use pigs to stir compost and till the garden, letting animals do the work. We're also using them in a forest-to-pasture conversion project that now covers nearly 2 acres up in the woods. We've reforested about 60 acres for conservation projects and to get healthy forestal ecosystems within 200 yards of all the open land and riparian areas. Now we've identified some flat areas that as pasture would complement the forest.

We clean off the trees, sawing the big ones into lumber for our building projects and the smaller material for firewood. Then we sow small grain and let the pigs harvest it. Last year we planted open-pollinated field corn and let the pigs hog it down. This year we'll plant more — including zucchini squash — to offer the pigs a salad bar. We move them every few days to fresh paddocks. The pork is spectacular. We call them real "bush hogs."

Like the beef, we do have pork in the freezer by the cut, but it is far more expensive than buying it by the half. We cured our own pork last year and it is excellent. We are trying to figure out how to offer this without getting in trouble with the inspection gestapo. We'll keep you posted. Prices are the same as last year.

PASTURED RABBIT This is 15-year-old Daniel's enterprise and he hopes to sell more than 1,000 this year. For fine dining pleasure, you can enjoy his rabbit at Joshua Wilton House in Harrisonburg, and the Metro-politan and Boar's Head Inn in Charlottesville. Remember that rabbit offers the highest protein of all meats without saturated fat because a rabbit only stores fat in its viscera, not in its muscle. He is increasing the price a little this year due to feed cost increases.

Last year he planted and harvested a bed of mangels — like sugar beets --

and fed them to the rabbits in order to cut feed costs and substitute our own homegrown goodies. They did extremely well and he hopes to grow much more this year.

PASTURED EGGS We supply several white tablecloth restaurants with these premium eggs. Raised on fresh pasture in portable housing, these birds lay eggs that were food lab tested and found to contain 3 times more Omega 3 and Omega 6 fatty acids than supermarket eggs. These are the marine oils, otherwise known as the cholesterol reducers. Greens in the diet, fresh air, sunlight and exercise completely change the nutritional profile of meat, poultry and eggs. Please save and bring your old egg cartons. We need tons of them. We're keeping the same price as last year.

PASTURED TURKEY Many of you had a more memorable Thanksgiving last year because of these delectable birds, raised just like the broilers in portable, floorless houses that protect them from weather and predators, but offer them a new salad bar pasture spot each day. We'd like you to try some throughout the season this year instead of just at Thanksgiving. Remember that if you want one for Christmas, you'll need to use your freezer.

Turkeys actually eat more grass and clover than chickens, producing a more profound difference between the taste and texture of our birds compared to factory confinement ones. Last year our season's average was about 15 pounds per bird and that seemed to suit most of you. Our processing equipment cannot handle 20 pounders.

We've had to go up a little in price because the poults cost nearly $2 and our labor requirements are higher than we thought. These birds require 16 weeks to grow, which means nearly 3 months of daily moves. We hope a couple more bucks a bird won't be too big a burden for you, but for us, multiplied by several hundred, it adds up to enough for our return-to-labor requirements.

VEGETABLES We usually have a few vegetables throughout the season, although this is not a large enterprise for us. Most of you have gardens because you are can-do people; that's wonderful. We'll try to let you know what is available when you come out.

STAUNTON/AUGUSTA FARMERS' MARKET Last year we added the farmers' market venue to our marketing, primarily to give our apprentices and our children firsthand sales experience. It allowed us to touch a new level of customer — the one who is wanting to opt out of supermarket fare, but is not yet purchasing directly off the farm. Although we were the only vendor with meat, the response was not as good as we had hoped. We had two main problems:

1. Being able to display our wares. Vegetables and baked goods do not require refrigeration and can be displayed easily. Customers can browse and get a full picture of what you have for sale. Our stuff was in a freezer on the back of the pickup truck, and not in plain view. We hope to do some things this season that will reduce this problem.

2. Limited offering. We ran out of beef in the first month and had nothing but ground beef for the lion's share of the season. That reduced us to just pork because due to market rules we could not sell our poultry, turkey and rabbit. The reduced inventory worked against us.

This will be solved this year because we have much more beef going into the season, plenty of pork and the inspection people in Richmond have approved our poultry and rabbit sales at the market. With a wry grin, the market master told us: "Everyone down in Richmond knew you."

Of course, our prices at the market are higher than here at the farm because there we must pay a market commission on gross sales. While we realize many folks will buy there and never come to the farm, our goal is still to see everyone come to the farm, to build a relationship with their food supply, and to experience a clean farm.

For the best in local, fresh, high quality vegetables and baked items, we encourage you to patronize the market every Saturday from 7 a.m. until noon beginning April 12. Rachel, the 10-year-old around Polyface, makes a wonderful pound cake and Daniel makes maple sugar donuts. He tapped the trees in the yard this year and made his own syrup. We revel in the next generation's entrepreneurial spirit.

OLD BUSINESS Last year, we mentioned a lobby organization called

Food Alternatives for Relationship Marketing to get inspection exemption for on-farm sales so we could sell a pound of sausage from a Thanksgiving hog killin' to a neighbor without a $200,000 processing plant. We have disbanded that organization but are keeping the torch lit with a similar effort through the state government. We corresponded with many of you this winter on a bill to allow on-farm sales and it was killed in committee, but we did get a joint study resolution, Senate Resolution No. 29, establishing a select committee to study on-farm sales of agricultural products.

Our state Senator Emmett Hanger sponsored the resolution in the senate and our Delegate Vance Wilkins sponsored it in the House. We spent several days in Richmond lobbying for this effort and owe a great debt to Christine Solem, commonly known as "the raw goat milk lady," for seeing this effort go as far as it did. We do not yet know when the hearings will be or where, but we will keep you posted. We desperately need to pack those hearing rooms with people who want to preserve freedom of choice in the food supply. If regulations continue the way they are, you'll be hard pressed to get anything besides industrial pseudo-food from the huge corporate "food" industry.

Another item is the cookbook. We thank those of you who sent recipes and still want to proceed with that project. The offer is still good: send us at least two recipes and we'll give you a free broiler this summer. We'll try to make sure this happens in the next year. Hurricane Fran put several projects on hold last fall.

OUTREACH This winter Joel spoke in various marketing and agriculture conferences in Texas, Arkansas, Iowa, Minnesota, Missouri, Pennsylvania, West Virginia, New York and Vermont. His two books, *PASTURED POULTRY PROFITS and SALAD BAR BEEF* are selling well and encouraging folks all over the country to try clean agriculture and relationship marketing. He brags on you all the time, encouraging farmers to find folks in their communities who want clean food and a link to the land. We want you to realize how big a part you are playing in the whole environmental foods movement. Thank you. Also, remember that Joel does slide programs for church groups, civic clubs, health organizations — whatever, about the clean food issue. If you're looking for an interesting program for your organization, give us a call and we can tailor it to your group.

FARM ACCESS You will be pleased to know that our bridge is fixed — better than before — after it was torn out by Hurricane Fran last September 5. We apologize for inconveniencing so many of you during the last couple batches of chickens and the Thanksgiving turkeys but we were in crisis mode for about three months. Thank you for your encouragement and understanding during one of the toughest times of our lives and the farm business. We are just now feeling like we're catching up to the lost time last fall. We plan to have a large sales building up by this summer to better handle our on-farm sales. We have been able to preserve some local heritage by purchasing the two counters from Augusta Frozen Food, which went out of business Jan. 1 this year. It will be big enough to accommodate all of our freezers and refrigerators, which will put everything in one place. We'll have storage for plastic bags and additional display space for crafts or baked items. In addition, we'll finally be able to have a literature rack in order to expose all of you to important information.

BROCHURE Enclosed please notice our first ever Polyface brochure. It is long overdue and we realize we've done this backwards. Normally a business does this kind of thing first, and then begins making product. But we've been so busy researching and developing clean food models that we've neglected the finer public relations endeavors. This is meant to introduce someone to us, a "first inquiry" type of thing. But we wanted to give each of you one to keep you abreast of our development. If you need a couple extras, let us know and we'll send them right out.

We keep a very tight customer list, so if your envelope has an "L" underneath your address, that means this will be your last newsletter and order form. We trim "inactives" each year. Thank you for taking the time to review this annual communiqué from your farm. We hope each of you feels a sense of partnership in this ministry, that you will each catch the vision of clean food, humane animal husbandry and agriculture with a conscience. Please know that our commitment to excellence is deeper than you can imagine, and we deeply appreciate — with due sobriety — your entrusting us with the integrity of your food. We can't wait to see you this season. God bless each of you.
THE SALATINS OF POLYFACE INC.

ORDER BLANK ***SPRING 1997*** ***POLYFACE, INC.***

Rt. 1, Box 281, Swoope, VA 24479 540-885-3590

PASTURED POULTRY

BROILERS $1.45/lb.	STEWING HENS $1.35/lb.	
Number of birds wanted		Processing dates
______	______	May 20,21,23, 24
______	______	June 17, 18,20,21
______	______	July 8,9,11,12
______	______	July 29,30, Aug 1,2
______	______	Aug 19,20,22,23
______	______	Sept 9,10,12,13
______	______	Sept 30, Oct 1,3,4

Chicken liver YES NO EXTRA

SALAD BAR BEEF

Late October. Prices are $1 per head plus shipping and handling based on carcass weight; includes cut, wrap and freeze. Take-home weights are 25-35 percent less than carcass hanging weight due to trimming and deboning. Ditto pork.

______	Split half	$1.95	85-150 lbs.
______	Half	$1.90	170-300 lbs.
______	Whole	$1.85	350-650 lbs.
______	Ground qtr.	$1.45	100-150 lbs.

______ Do you want liver?

NOTE: Circle weights if you strongly prefer large or small.

PASTURED RABBIT

Available year-round $3.00/lb.

_______ How many?

Roughly, when? ______________________

____

____

PIGAERATOR PORK

Prices and policy exactly like beef.

June-July		October	
_______	Half $2.05 (100-150 lbs.)	_______	Half
_______	Whole $2.00 (200-300 lbs.)	_______	Whole

READY-TO-LAY PULLETS

$5.75 apiece

June ______________ How many?

Sept. ______________ How many?

Primarily Rhode Island Reds, but also White Rocks, Barred Rocks and Black Australorps

PASTURED TURKEY $1.75/lb.

How many?	Weeks available
___________	July 29
___________	Aug. 19
___________	Sept. 9
___________	Sept. 30
___________	Thanksgiving

NAME ____________________________ PHONE (________) __________________

ADDRESS ____

Comments, questions, suggestions welcome on back. Document this order and post on the refrigerator so you won't forget what you did.

Resources

Periodicals

ACRES USA
P.O. Box 8800
Metairie, LA 70011
(504) 889-2100

APPPA *Grit!*
American Pastured Poultry
Producers' Association
5207 70th Street
Chippewa Falls, WI 54729
(715) 723-2293

Holistic Management Quarterly
Albuquerque, NM 87102
(505) 842-5252

Quit You Like Men
P.O. Box 1050
Ripley, MS 38663-9430
(601) 837-4596

Small Farmer's Journal
P.O. Box 1627
Sisters, OR 97759
(503) 549-2064

American Small Farm
9420 Topanga Canyon
Chatsworth, CA 91311-5759
(818) 727-2236

Countryside and Small Stock Journal
N2601 Winter Sports Rd.
1007 Luna Circle NW
Withee, WI 54498
(800) 551-5691

Permaculture Activist
P.O. Box 1209
Black Mountain, NC 28711
(704) 683-4946

Small Farm Today
3903 W. Ridge Trail Rd.
Clark, MO 65243-9525
(800) 633-2535

Stockman Grass Farmer
5135 Galaxie Drive
Suite 300C
Jackson, MS 39206
(800) 748-9808

Books

Bartholomew, Mel
Square Foot Gardening

Beckwith, Harry
Selling the Invisible

Belanger, Jerome D.
Raising Small Livestock

Berry, Wendell
The Gift of Good Land
The Unsettling of America

Bromfield, Louis
Out of the Earth
Pleasant Valley

Byczynski, Lynn
The Flower Farmer: An Organic Grower's Guide to R aising and Selling Cut Flowers

Celente, Gerald
Trends 2000

Cobleigh, Rolfe
Handy Farm Devices and How to Make Them

Coleman, Eliot
Four-Seasons Harvest
The New Organic Grower

Cook, Glen Charles
500 More Things to Make for Farm and Home

Damerow, Gail
Chicken Health Handbook

Davidson, James Dale
The Sovereign Individual

Davis, Adelle
Let's Cook it Right

Doane, D. Howard
Vertical Diversification

Douglas, J. Sholto
Forest Farming

Drucker, Peter F.
Managing in a Time of Great Change

Ensminger, M. E.
Beef Cattle Science

Faulkner, Edward H.
A Second Look
Plowman's Folly
Soil Development

Ferguson, John
Farm Forestry

Forbes, Reginald
Woodlands for Profit and Pleasure

Fryer, Lee
The Bio-Gardener's Bible

Gibson, Eric
Sell What You Sow

Groh, Trauger
Farms of Tomorrow

Hensel, Julius
Bread from Stones

Howard, Sir Albert
An Agricultural Testament

Ishee, Jeff
Dynamic Farmers' Marketing

Jackson, Wes
Altars of Unhewn Stone

Leatherbarrow, Margaret
Gold in the Grass

Lee, Andy
Backyard Market Gardening
Chicken Tractor

Levinson, Jay Conrad
Guerrilla Marketing Attack

Lindegger, Max O.
The Best of Permaculture

Logsdon, Gene
The Contrary Farmer
Successful Berry Growing

Mollison, Bill
Permaculture: A Designer's Manual
Permaculture One
Permaculture Two

Morrison, Frank B.
Feeds and Feeding

Morrow, Rosemary
Earth User's Guide to Permaculture

Murphy, Bill
Greener Pastures on Your Side of the Fence

Nation, Allan
Grass Farmers
Pasture Profits with Stocker Cattle

Popcorn, Faith
Clicking: 17 Trends That Drive Your Business—and Your Life

Rodale (staff)
Complete Book of Composting
Encyclopedia of Organic Garden ing
How to Grow Vegetables and Fruits by the Organic Method
Rodale Book of Composting

Salatin, Joel
Pastured Poultry Profits
Salad Bar Beef

Savory, Allan
Holistic Resource Management

Siegmund, Otto H., Editor
The Merck Veterinary Manual

Smith, Burt
Intensive Grazing Management: Forage, Animals, Men, Profits

Smith, Russell
Tree Crops: A Permanent Agriculture

Staten, Hi W.
Grasses and Grassland Farming

Thompson, W. R.
The Pasture Book

Turner, Newman
Fertility Pastures and Cover Crops

Voisin, André
Grass Productivity

Walters, Charles, Jr.
An ACRES USA Primer
Weeds, Control Without Poisons

Whatley, Booker T.
How to Make $100,000 Farming 25 Acres

Whitefield, Patrick
How to Make a Forest Garden

Willis, Harold
The Coming Revolution in Agriculture

Yeomans, P. A.
Water for Every Farm: Yeomans Keyline Plan

Catalogues

BASS EQUIPMENT COMPANY
1-800-798-0150
Small animal cage materials and hardware, specializing in rabbits

CHUCK WAGON OUTFITTERS
1-800-543-2359
Cast iron cookware and accessories

DIRECT LINE
1-800-241-2197
Power tools, generators, pumps, air tools, bearings and parts

EDIBLE LANDSCAPING
1-804-361-9134
Located in Afton, Virginia, catalogue offers host of perennial fruit and nut varieties.

FORESTRY SUPPLIERS, INC.
1-800-543-4203
Probably THE leader in forestry equipment and tools, but also offers sprayers, hoses, fertilizer spreaders for 4-wheelers, diagnostic and soil testing kits, protective clothing, surveyng equipment, compost thermometers, environmental science and engineering — one of my personal favorites.

GEMPLER'S
1-800-382-8473
Protective outerwear like boots, rain gear; safety devices, step ladders and first aid kits

GRAINGER
1-540-982-3543
More than 78,000 products, 350 branches, offering parts, tools, welding supplies, motors and accessories, gears, pulleys

HOOBER
1-717-768-8231
Farm equipment, parts, filters, shop accessories

HOEGGER
1-800-221-4628
Goat and cheesemaking supplies, including butter churns

JEFFERS
1-800-533-3377
Livestock medical supplies, tools, ear tags and electric fencing

LEHMAN'S NON-ELECTRIC
1-330-857-5757
Especially for folks without electricity, items for simple, self-sufficient living

LIONEL INDUSTRIES
1-561-624-9093
Egg and poultry supplies

NASCO FARM AND RANCH
1-800-558-9595
Livestock supplies and equipment, cattle scales, boots, traps, testing kits

NORTHERN GREENHOUSE SALES
1-204-327-5540
Woven poly for greenhouses, hoophouses. Catalogue is full of do-it-yourself and poor-boy ideas.

NORTHERN HYDRAULICS
1-800-556-7885
Power tools, casters, meters, small engines

RAND
1-800-755-7263
Materials handling and packaging products, wheel conveyers, affordable electronic scales and wrapping materials

RELIABLE
1-800-735-4000
Office supply products

RUBBERMAID AND MATTING CATALOG
1-800-362-1000
Every conceivable container and matting material

SETON
1-800-243-6624
Identification products, packaging, signs and decals

SEVEN SPRINGS FARM
1-540-651-3228
Pest management products, animal supplements, natural fertilizers

SMITH AND HAWKEN
1-800-776-3336
Garden tools and accessories — extremely high quality

SNOW POND FARM SUPPLY
1-800-768-9998
Soil amendments, season extenders, tools and supplies, pest controls and seed potatoes

STUPPY'S
1-800-877-5025
Greenhouses and accessories

TEK SUPPLY
1-800-835-7877
Floodlights, overhead doors, industrial mats, alarms, squeegees, material regulators,fans, control timer devices

WORM'S WAY
1-800-274-9676
Organic soil amendments and biologicals, season extenders and seeds

Index

A

C

E

F

G

H

I

J

K

L

M

N

O

P

Q

R

S

T

U

V

W

Y

Z